AS UNIT 2

STUDENT GUIDE

CCEA

Physics

Waves, photons and astronomy

Ferguson Cosgrove

HODDER
EDUCATION
AN HACHETTE UK COMPANY

Hodder Education, an Hachette UK company, Carmelite House, 50 Victoria Embankment, London, EC4Y 0DZ

Orders

Hachette UK Distribution, Hely Hutchinson Centre, Milton Road, Didcot, Oxfordshire, OX11 7HH

tel: 01235 827827

e-mail: education@hachette.co.uk

Lines are open 9.00 a.m.–5.00 p.m., Monday to Friday. You can also order through the Hodder Education website: www.hoddereducation.co.uk

ISBN 978-1-4718-6393-6

First printed 2016

Impression number 7

Year 2022

This guide has been written specifically to support students preparing for the CCEA AS and A-level Physics examinations. The content has been neither approved nor endorsed by CCEA and remains the sole responsibility of the author.

Cover photo: kasiastock/Fotolia

Typeset by Integra Software Services Pvt. Ltd, Pondicherry, India

Printed and bound by CPI Group (UK) Ltd, Croydon, CR0 4YY

Hachette UK's policy is to use papers that are natural, renewable and recyclable products and made from wood grown in well-managed forests and other controlled sources. The logging and manufacturing processes are expected to conform to the environmental regulations of the country of origin.

Contents

■ Getting the most from this book

Sample student answers

Practise the questions, then look at the student answers that follow.

Exam-style questions

Commentary on the questions

Tips on what you need to do to gain full marks, indicated by the icon ⓔ

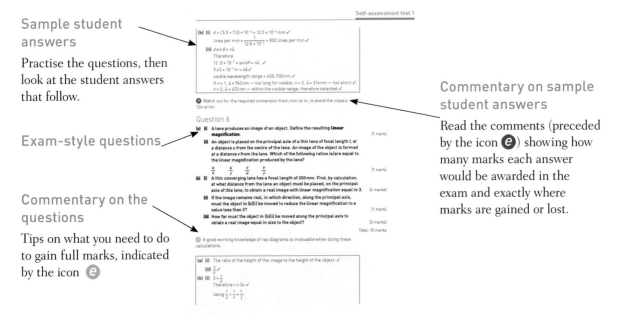

Commentary on sample student answers

Read the comments (preceded by the icon ⓔ) showing how many marks each answer would be awarded in the exam and exactly where marks are gained or lost.

■About this book

This guide is one of a series covering the CCEA specification for AS and A2 physics. It offers advice for the effective revision of Unit AS 2: Waves, photons and astronomy. Its aim is to help you understand the physics and give you guidance on the core aspects of the subject. The guide has two sections:

- The **Content Guidance** covers Unit AS 2. It does not have the detail of a textbook but it offers guidance on the main areas of the content and includes worked examples. These examples illustrate the types of question that you are likely to come across in the examination. The exam tips will help you to understand the physics and give you guidance on the core aspects of the subject. They also show how to approach revision and improve your exam technique.
- The **Questions & Answers** section comprises two self-assessment tests. Answers are provided and there are comments on the specific points for which marks are awarded.

Physics is not an easy subject, but by committing time and effort to understanding the key elements of the discipline you can maximise your performance in the examination. The development of an understanding of physics can only evolve with experience and practice. This guide will facilitate your progress by focusing on the essential components and providing examples for you to attempt before learning from the answers.

The specification

The CCEA specification is a detailed statement of the physics that is required for the unit assessments and describes the format of the assessments. It can be obtained from the CCEA website at www.rewardinglearning.org.uk.

Your teacher may introduce you to concepts outside the specification to further develop your physics.

Revision tips

- Be familiar with the specification.
- Organise your notes and make sure they are complete.
- Learn all the equations indicated in the specification as those you must recall.
- Be familiar with the equations that are provided on the *Data and Formulae Sheet*.
- Practise rearranging and using equations to find different quantities.
- Learn definitions and laws thoroughly and accurately.
- Be able to describe all the experiments referred to in the specification with the aid of a labelled diagram.

Content Guidance

■ Waves

- A wave is a disturbance that propagates through a medium.
- Waves carry energy.
- There are many ways to classify waves.

Progressive and standing waves

Progressive waves transfer energy from one place to another. They are sometimes called travelling waves as they appear to move. Examples of progressive waves are electromagnetic waves travelling from the Sun to Earth or a sound wave travelling from a loudspeaker to an ear.

Standing waves do not involve the transmission of energy, as the wave energy is stored in the system. They are sometimes called stationary waves as they do not appear to move. Examples of standing waves are the vibrations in a violin string or air vibrating inside a flute.

Mechanical and electromagnetic waves

Mechanical waves are produced by a disturbance in a material and are transmitted by the oscillating particles of the material. Examples of mechanical waves are water waves or waves on a slinky spring

Electromagnetic waves consist of varying electric and magnetic fields. Examples include any member of the electromagnetic spectrum.

Transverse and longitudinal waves

A **transverse wave** is one in which the vibrations of the particles (or the variations in the electric and magnetic fields) are at right angles to the direction in which the wave travels. Examples are surface water waves or electromagnetic waves.

A **longitudinal wave** is one in which the direction of the vibrations of the particles is parallel to the direction in which the wave travels. Examples include sound waves. The back-and-forth oscillations of the air particles forms regions of high and low pressure called **compressions** and **rarefactions** respectively.

Longitudinal and transverse waves can be demonstrated using a slinky spring (Figure 1a and b).

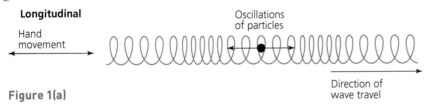

Longitudinal

Hand movement

Oscillations of particles

Direction of wave travel

Figure 1(a)

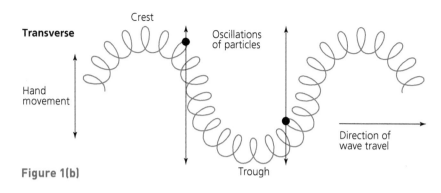

Knowledge check 1

Distinguish between
the nature of the
vibrations in transverse
and longitudinal waves
and give **two** examples
of each type of wave.

Transverse

Crest

Oscillations
of particles

Hand
movement

Direction of
wave travel

Trough

Figure 1(b)

Graphical representation of waves

It is useful to be able to represent all types of wave graphically. It is conventional to
consider the case of regular, repeated disturbances setting up the wave. Two different
graphs are possible:

- A displacement–distance graph is effectively a snapshot of a section of the wave,
  showing the displacement of all the particles at an instance in time.
- A displacement–time graph shows how the displacement of one particle in the path
  of the wave changes over an interval of time.

The displacement–distance graph

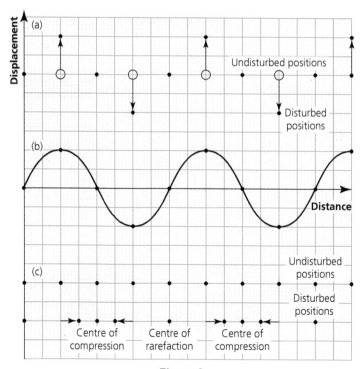

Figure 2

For transverse waves the actual view of the wave and the graphical representation are identical, as shown by Figures 2(a) and 2(b). It is not so obvious in the longitudinal case, because the longitudinal displacement must be, in effect, turned through 90° and plotted on the vertical axis, as shown in Figures 2(c) and 2(b). So the sinusoidal waveform shown in Figure 2(b) is valid for both wave types.

The displacement–time graph

Here we consider the displacement — transverse or longitudinal — of one particle in the medium through which the wave is passing over a time interval. The repeated up–down or right–left movement will again result in a sinusoidal graph, as shown in Figure 3.

Exam tip

It is normal to assign up or right as the positive displacement and down or left as negative.

Wave properties

Now that we have shown that the oscillating particles of a wave can be represented on two types of graph, it is possible to relate the properties of a wave directly.

■ The displacement–time graph for a single particle within the wave shows how the displacement of this particle from its equilibrium position varies with time (Figure 3).

Exam tip

It is easy to confuse these graphs. Always check which graph you are using by looking carefully at the labels on the axes.

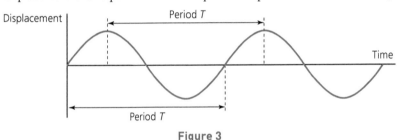

Figure 3

■ A displacement–distance graph shows the position of all the particles in a section of the wave at a single instant (Figure 4).

Knowledge check 2

a How can frequency be determined from a displacement–time graph? **b** Can frequency be determined from a displacement–distance graph?

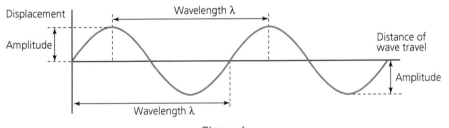

Figure 4

Definitions

Now that we have visual representations, it is easier to understand the definitions.

■ The **displacement** (transverse or longitudinal), x, of a particle on a wave is its distance from the undisturbed position of the oscillation.

■ The **amplitude**, A, of the wave is the maximum displacement of a particle from the undisturbed position of the oscillation. The unit is the metre (m).

Exam tip

Care should be taken to distinguish between displacement and amplitude. Amplitude is the *maximum* displacement.

- A **crest** is a point on the wave of maximum displacement in the positive direction.
- A **trough** is a point on the wave of maximum displacement in the negative direction.
- The **periodic time**, T, or **period**, is the time taken for one complete oscillation of the wave. The unit is the second (s).
- The **frequency**, f, is the number of complete waves that pass a point in one second, or the number of oscillations per second. The unit is the hertz (Hz).
- The **wavelength**, λ, of the wave is the distance the waveform progresses in the periodic time. The unit is the metre (m).
- The **wave speed**, v, is the distance travelled by the wave each second. The unit is metres per second (m s^{-1}).

> **Exam tip**
>
> The **wavelength**, λ, is *numerically equal to* the distance between consecutive points of corresponding phase. It can be measured as the distance between two adjacent crests or between two adjacent troughs. (But this is not the definition of the quantity.)

The wave equation

The **wave equation** relates the speed of a wave to its frequency and wavelength. By definition, a wave travels one wavelength, λ, in the time taken for one oscillation, T.

As speed is the distance moved per unit time:

$$v = \frac{s}{T} \text{ and } v_{\text{wave}} = \frac{\lambda}{T} = \frac{1}{T} \times \lambda = f \times \lambda$$

The wave equation is:

velocity = frequency × wavelength

$$v = f\lambda$$

where v is the speed of the wave in metres per second (m s^{-1}), f is the frequency of the wave in hertz (Hz) and λ is the wavelength in metres (m).

> **Exam tip**
>
> Always check that the units are correct and consistent. Speed and wavelength must use the same units of length. Questions involving the wave equation often involve units with prefixes such as MHz, kHz, μm and nm. Learn all the prefixes for multiples and sub-multiples of units and practise putting them into your calculator.

Worked example

A girl generates a wave on a rope by moving her hand up and down, as shown in Figure 5. She generates two waves every second.

a What three types of wave is she generating?

b What is the amplitude of the waves?

> **Exam tip**
>
> Periodic time and frequency are related. A frequency of 10 Hz means 10 oscillations per second or one oscillation is completed in 1/10 second. So the periodic time of this oscillation is 0.1 s.
>
> $$\text{frequency} = \frac{1}{\text{period}}$$

> **Knowledge check 3**
>
> Calculate the frequency of a wave with a periodic time of 1 ms.

> **Exam tip**
>
> Frequency and wavelength are inversely proportional. Lower-frequency waves have longer wavelengths. Higher-frequency waves have shorter wavelengths.

> **Exam tip**
>
> Always state units correctly using a negative index, not a slash. For example, the units of velocity are m s^{-1} *not* m/s.

→

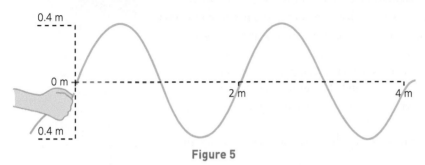

Figure 5

c What is the frequency of the waves?

d What is the speed of the waves?

Answer

a Transverse waves, as the vibrations are at right angles to the direction of the wave.

Mechanical waves, as the energy is transmitted by the oscillating particles of the rope.

Progressive waves, as the energy is being transferred from one place to another.

b 0.4 m (amplitude is the maximum displacement from the mid-point of the wave)

c Frequency = 2 waves per second = 2 Hz (frequency is the number of waves per second)

d $\lambda = 2\,m$ (wavelength is the distance between two crests)

$v = f \times \lambda = 2 \times 2 = 4\,m\,s^{-1}$

Phase and phase difference

Phase describes the particular point in the cycle of a wave. It is used to compare the motion of vibrating particles in a wave or waves. Consider the motion of a single particle with the displacement–time graph in Figure 6.

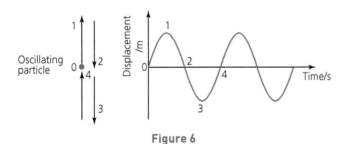

Figure 6

The points labelled 1, 2, 3 and 4 represent the position of the particle at different stages of one complete cycle of the oscillation. One full cycle or one oscillation of a wave (from point 0 to point 4) is considered to be 360° or 2π radians.

■ The timing of the oscillation begins at point 0 when the particle is passing up through the mid-point of the oscillation.

■ At point 1, the particle is at the positive maximum displacement and has completed one quarter of the cycle of the oscillation. This stage is out of phase with point 0 as

Exam tip

When answering questions involving calculations follow this four-point plan:

1 Write down the **formula** you are going to use and change the subject of the formula to the unknown.

2 **Substitute** the quantities into the amended formula, making sure that you have consistent units for the same quantities.

3 **Calculate** the answer.

4 Include the correct **unit** with your answer.

Knowledge check 4

A bat emits a sound pulse of wavelength 5.0 mm and frequency 68 kHz. Calculate the speed of the emitted sound.

it is different by one quarter of the cycle of the oscillation. It is 90° or π/2 radians out of phase with point 0.

■ At point 2, the particle is moving down through the mid-point of the oscillation. This is the same displacement as at point 0, but not moving in the same direction. It has completed half of the oscillation. It is 180° or π radians out of phase with point 0. This is known as antiphase.

■ At point 3, the particle is at the negative maximum displacement and has completed three-quarters of the oscillation. It is 270° or 3π/2 out of phase with point 0.

■ At point 4, the particle has completed one oscillation and is at the same position and moving in the same direction as at point 0. It is in phase with point 0.

Consider the motion of the particles in a wave in the displacement–distance graph in Figure 7. Now we are comparing the motion of different particles at a single instant along a section of a wave. Particle A is in phase with particle E, as it is the same displacement from the mid-point of its oscillation and moving in the same direction. Particle B is in antiphase with particle D as it is π radians out of phase, but it is in phase with particle F.

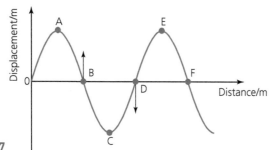

Figure 7

We can also use phase to compare two waves with the same frequency (Figure 8). One wave is leading the other wave, or one wave is lagging behind the other.

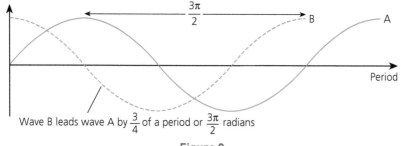

Wave B leads wave A by $\frac{3}{4}$ of a period or $\frac{3\pi}{2}$ radians

Figure 8

The phase difference is calculated between any two waves of the same frequency by finding the fraction of the complete 2π radians that represents the difference in phase. The two waves are out of step by a time t. The phase difference is equal to $(t/T) \times 2\pi$ or $(t/T) \times 360°$.

Exam tip

One crest and one trough together is one whole wave. This can be described in terms of angles. One wave is 360° or 2π radians.

Knowledge check 5

A wave of period 14 s lags behind another wave of the same frequency by 3.5 s. State the phase difference of the waves.

Worked example

For each of the graphs in Figure 9 state the phase difference in words, in terms of wavelength, in degrees and in radians.

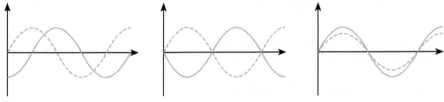

Figure 9

Answer

a phase difference = 90° or $\lambda/4$ or $\pi/2$ — out of phase

b phase difference = 180° or $\lambda/2$ or π — completely out of phase

c phase difference = 360° or λ or 2π — in phase

Polarisation

The condition for a wave to be plane-polarised is for the oscillations to be in just one plane. This phenomenon only occurs with transverse waves, for example light waves.

In normal, unpolarised light waves, such as light from a filament lamp, electric field oscillations occur in all planes perpendicular to the direction of travel of the wave. If this unpolarised light is passed through a polarising filter (called Polaroid), the oscillations in all planes but one will be absorbed. The light emerging from the filter has oscillations in one plane only and is said to be plane-polarised.

As only one plane is now transmitted, the intensity of the light is substantially reduced.

A second sheet of Polaroid can be used to confirm that the light is plane-polarised. The second sheet is called the analyser and it is held in line with the polariser. The analyser is rotated through 360° and the light intensity will alternate between maximum and minimum (extinction) light intensities every 90°. The intensity depends on whether the transmission axis of the analyser is parallel or perpendicular to the transmission axis of the polariser (Figure 10).

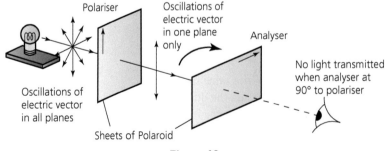

Figure 10

Note: longitudinal waves cannot be plane-polarised as the oscillations are parallel to the direction of wave travel.

Checking that a wave is polarised

An analyser is used to check if waves are polarised. An analyser produces a polarised wave itself. If the original beam is unpolarised, as the analyser is rotated it will continually polarise the waves in successive planes as it is rotated. There will always be one plane that is transmitted. Hence the intensity remains constant (Figure 11).

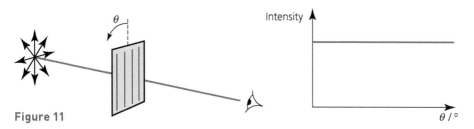

Figure 11

If the original beam is polarised, as the analyser is rotated the intensity will vary from a maximum when the analyser is aligned with the plane of polarisation to a minimum when the analyser is at right angles to the plane of polarisation (Figure 12).

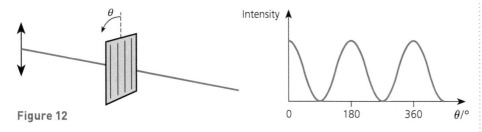

Figure 12

Worked example

A microwave generator produces plane-polarised microwaves. An aerial connected to an amplifier and ammeter can be used to analyse the polarisation of the waves. When the aerial is parallel to the plane of polarisation, the maximum signal is received. As the aerial is rotated the signal intensity reduces, reaching a minimum when the aerial has rotated through 90°.

a What is meant by plane-polarised microwaves?

b Draw a labelled diagram of the apparatus you would use to demonstrate that the microwaves are plane-polarised.

c What does this experiment demonstrate about the nature of microwaves?

d Sketch a graph to show how the intensity of the signal changes as the aerial is rotated.

e Explain why sound waves cannot be plane-polarised.

Answer

a Plane-polarised microwaves are ones in which the oscillations are confined to one plane only, perpendicular to the direction of the wave travel.

➡

b

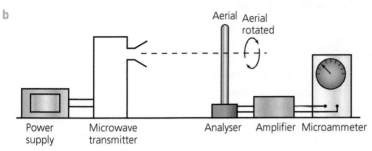

c As microwaves are plane-polarised the experiment shows they are transverse waves.

d

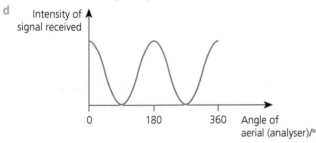

e Sound is a longitudinal wave and longitudinal waves cannot be plane-polarised as the oscillations are parallel to the direction of wave travel.

Electromagnetic waves

Electromagnetic waves cover a large range of wavelengths called the electromagnetic spectrum. This continuous spectrum is divided into regions of waves depending on how the waves are produced. These broad regions overlap (Table 1) — there is no clear boundary between regions.

All electromagnetic waves:

- are transverse waves
- consist of varying electric and magnetic fields
- can travel through a vacuum
- travel at a speed of $3.00 \times 10^8 \, \mathrm{m \, s^{-1}}$ in a vacuum

Table 1 Electromagnetic waves

Region of the electromagnetic spectrum	Typical wavelength range/m	Applications
Gamma rays	10^{-16} to 10^{-12}	Sterilising medical equipment Medical imaging — CT scans Killing cancerous cells
X-rays	10^{-12} to 10^{-9}	Airport security scanners Medical Imaging — detecting broken bones Studying crystal structures
Ultraviolet	10^{-9} to 10^{-7}	Vitamin D production in the body Sun beds Causes fluorescence in security marking and detecting forged bank notes

Region of the electromagnetic spectrum	Typical wavelength range/m	Applications
Visible light	4×10^{-7} (violet) to 7×10^{-7} (red)	Photosynthesis Photography Vision
Infrared	10^{-6} to 10^{-4}	Cooking, grilling Remote controls Thermal imaging devices
Microwaves	10^{-4} to 10^{-1}	Cooking Mobile phones Satellite communications
Radio	10^{-1} to 10^5	Radio and television communications Radio astronomy

Exam tip

Learn the order of the electromagnetic spectrum and know which is the high-frequency end and which is the long-wavelength end.

Knowledge check 7

State **three** properties of radio waves and explain why they are not considered as being dangerous.

Worked example

The speed of electromagnetic waves in air is given by $c = 3 \times 10^8 \, \text{m s}^{-1}$. Calculate:

a the wavelength of radio waves of frequency 250 kHz

b the frequency of microwaves of wavelength 5 cm

Answer

a $v = f\lambda \Rightarrow \lambda = \dfrac{v}{f} = \dfrac{c}{f} = \dfrac{3 \times 10^8}{2.5 \times 10^5} = 1200 \, \text{m}$

b $v = f\lambda \Rightarrow f = \dfrac{v}{\lambda} = \dfrac{c}{\lambda} = \dfrac{3 \times 10^8}{0.05} = 6 \times 10^9 \, \text{Hz}$

Exam tip

Use the correct units in the formula. Convert frequency to hertz. Convert wavelengths to metres. A common mistake is to use kHz or cm.

Summary

- Waves can be classified as transverse or longitudinal. In transverse waves, the vibrations are perpendicular to the direction of wave travel. In longitudinal waves, the vibrations are parallel to the direction of wave travel.
- There are certain characteristics that describe a wave. The displacement of a particle is the distance from the undisturbed point of the oscillation and the amplitude is the maximum displacement. The periodic time is the time taken for one complete oscillation of the wave. The frequency is the number of complete waves that pass a point in one second. The wavelength is numerically equal to the distance between consecutive points of corresponding phase.
- The speed of a wave is the product of its frequency and wavelength, $v = f\lambda$. This is called the wave equation. Periodic time is related to frequency by $T = 1/f$.

- Graphs can be plotted to show the nature of the wave. The displacement–time graph shows how the displacement of a particle varies with time. Periodic time can be determined from this graph. The displacement–distance graph shows the position of all particles in a section of the wave at a single instant. Wavelength can be determined from this graph.
- Phase describes the particular point in the cycle of a wave. Phase difference can be used to compare different points in a wave or points in different waves. Points are in phase if separated by a whole number of complete cycles of the wave. Points are completely out of phase if separated by an odd number of half cycles of the wave.
- A wave is plane-polarised when the oscillations occur in one plane only. Transverse waves can be polarised. Light and all the electromagnetic waves

➜

can be polarised. Longitudinal waves, for example sound, cannot be polarised. A polariser is used to polarise waves, such as Polaroid film for light. An analyser can be used to test for polarisation, for example another piece of Polaroid film for polarised light. As the analyser is rotated the light intensity varies and is extinguished twice in a 360° rotation.

■ The electromagnetic spectrum is a map of all types of electromagnetic radiation. It includes visible light and other types that the eye cannot detect. The spectrum can be listed in order of increasing wavelength or frequency. The most energetic waves have the highest frequency and shortest wavelength. The electromagnetic spectrum in order of increasing frequency is: radio waves, microwaves, infrared radiation, visible light (red–violet), ultraviolet radiation, X-rays, gamma rays. For electromagnetic waves the speed of the electromagnetic waves in a vacuum is $c = 3 \times 10^8\,\mathrm{m\,s^{-1}}$. So the wave equation becomes $c = f\lambda$ for electromagnetic waves.

■ Refraction

Waves travel at different speeds in different media. For example, sound travels at approximately $340\,\mathrm{m\,s^{-1}}$ in air, $1500\,\mathrm{m\,s^{-1}}$ in water and $5000\,\mathrm{m\,s^{-1}}$ in steel.

Refraction occurs when a wave *changes direction* on travelling between different media due to a change in speed. All waves can be refracted. For example, as surface water waves move from deep to shallow water there may be a change of direction (Figure 13).

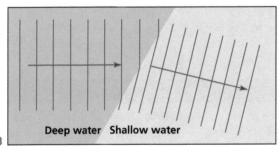

Figure 13

Deep water   Shallow water

If a wave slows down as it passes from one medium to the next, it will change direction towards the normal (Figure 14a). If the wave speeds up as it passes from one medium to the next, it will change direction away from the normal (Figure 14b). If the wave meets the boundary between the two media along the normal, at right angles to the surface, then the wave will change speed but there will be no change of direction and no refraction (Figure 14c).

Exam tip

In all diagrams showing refraction draw a normal, as all angles are measured from the normal. The normal is a line drawn at 90° to the boundary at the point of contact between the ray and the boundary.

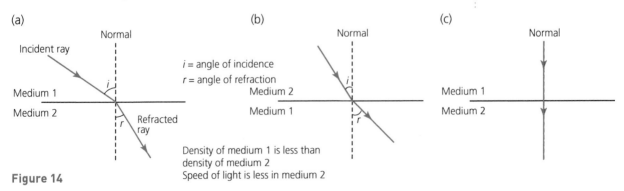

Figure 14

(a)
Normal
Incident ray
Medium 1
Medium 2
i
r Refracted ray

(b)
i = angle of incidence
r = angle of refraction
Normal
Medium 2
Medium 1
i
r
Density of medium 1 is less than density of medium 2
Speed of light is less in medium 2

(c)
Normal
Medium 1
Medium 2

Refraction can be shown by a ray of light passing through a parallel-sided glass block (Figure 15). Note that the emergent ray is parallel to the incident ray.

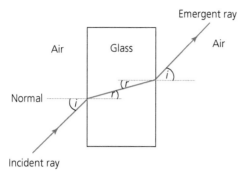

Figure 15

Explanation of refraction

Light travels at $3 \times 10^8\,\mathrm{m\,s^{-1}}$ in air. When it enters a denser material, it slows down. If the whole ray hits the new material at the same time, the whole ray slows down together and no change of direction occurs. However, if the ray strikes the surface at an angle, then some of the ray slows down before the rest. This leads to a change of direction, i.e. refraction (Figure 16).

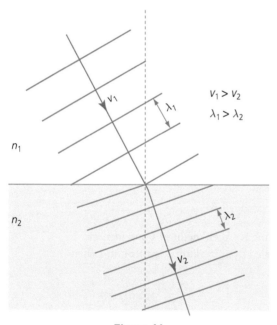

Figure 16

Exam tip

The change of direction associated with refraction is often loosely referred to as bending.

Exam tip

Refraction is caused by a change in wave speed. Refraction causes a change in wavelength. The frequency of the wave stays constant. In reflection and diffraction the wavelength stays the same.

Exam tip

Learn the wavefront diagram to explain how refraction occurs

Knowledge check 8

Describe changes, if any, to the speed, frequency and wavelength of water waves as they are refracted in going from shallow to deep water.

Refractive index

Refraction occurs at the boundary between two media because of the change of speed of the wave on entering the second medium. The ratio of the speed of the wave in the first medium v_1 to the speed of the wave v_2 in the second medium is called the **refractive index** for the boundary between those two media. It is denoted by $_1n_2$:

$$_1n_2 = \frac{\text{speed in medium 1}}{\text{speed in medium 2}} = \frac{v_1}{v_2}$$

Worked example

The refractive index of glass is 1.5. Calculate the speed of electromagnetic waves in glass.

Answer

$$_{air}n_{glass} = 1.5 = \frac{c_{air}}{c_{glass}}$$

$$c_{air} = 3.0 \times 10^8 \, \text{m s}^{-1}$$

$$c_{glass} = \frac{c_{air}}{1.5} = 2.0 \times 10^8 \, \text{m s}^{-1}$$

Snell's law

Snell's law states that the ratio of the sine of the angle of incidence to the sine of the angle of refraction is the same for all rays travelling across a given boundary.

This means that the ratio $\sin i / \sin r$ is a constant for any two given media. This ratio for the boundary between any two particular media is also equal to the refractive index.

Snell's law is very useful as it enables us to predict where a ray of light will go when it enters a medium.

Note that $_1n_2$ is the relative refractive index for light travelling between medium 1 and medium 2. If medium 1 is a vacuum then the refractive index is called the absolute refractive index. The absolute refractive indices of a pair of media determine their relative refractive index:

$$_1n_2 = \frac{n_2}{n_1} = \frac{\sin i}{\sin r}$$

$_1n_2$ is the relative refractive index for light travelling between medium 1 and medium 2. If the path of the light were reversed and it travelled from medium 2 to medium 1 then the refractive index would be written as $_2n_1$:

$$_1n_2 = \frac{1}{_2n_1}$$

Experiment to verify Snell's law

Apparatus: drawing board, paper, glass block, protractor, ruler, ray box

Draw the outline of the glass block and remove the block. Place an X at a point a third of the way along one of the longer sides. Construct a normal (perpendicular) at this point. Draw a line that intersects the block at X at an angle of 20° to the normal. Replace the block. Shine a ray of light along the line and mark the ray that emerges

➡

> **Exam tip**
>
> The definition of Snell's law must be stated exactly. Some students leave out a reference to a particular boundary and are deducted marks. The ratio will not work for the specific case of $i = r = 0°$, when light is *not* refracted. Note that there is no actual mention of refractive index in the law.

from the other side of the block. Remove the block and complete the lines to show the incident, refracted and emergent rays. The path of the ray through the block is complete. Measure the angle of refraction in the glass and record the result. Repeat the procedure for angles of incidence of 30°, 40°, 50° and 60° and record the results. Find the sine of the angles of incidence and the sine of the angles of refraction.

The results are recorded in a table such as Table 2.

Table 2 Recording Snell's law results

Angle of incidence $i/°$	Angle of refraction $r/°$	Sin i	Sin r
20.0	13.5	0.34	0.23
30.0	20.5	0.50	0.35
40.0	27.0	0.64	0.45
49.0	33.5	0.75	0.55
58.0	38.0	0.85	0.62

Snell's law gives

$$n = \frac{\sin i}{\sin r}$$

Rearranging

$$\sin i = n \sin r$$

Comparing with the straight line equation

$$y = mx + c$$

If a graph is plotted of $\sin i$ on the y-axis against $\sin r$ on the x-axis, a straight-line graph through the origin will verify Snell's law (Figure 17). The gradient of the straight line is equal to the refractive index of glass n.

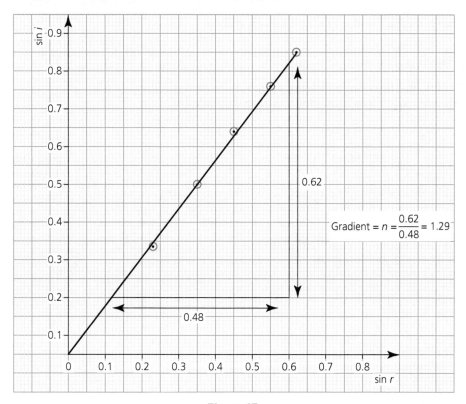

$$\text{Gradient} = n = \frac{0.62}{0.48} = 1.29$$

Figure 17

Exam tip

If a question requires you to describe an experiment, draw a labelled diagram with a ruler, include a procedure and how you will use the results to form a conclusion.

Exam tip

Results tables must have units. The units must follow the quantity at the top of each column and a slash (not brackets), indicating division, must be used to separate the quantity from the corresponding unit.

Exam tip

To confirm which graph to plot, rearrange the formula and do a direct mapping to the equation of a straight line, $y = mx + c$ (m is the gradient and c is the y-intercept).

Worked example

Light is incident on a layer of oil on the surface of a tank of water at an angle of 57°. If the refractive index of oil is 1.26 and that of water is 1.3 calculate:

a the angle of refraction in the oil

b the angle of refraction in the water

c the angle of refraction in the water if the oil is removed

d Comment on the answers to parts (b) and (c).

Answer

a angle of incidence = $i = 57°$

$$\frac{\sin i}{\sin r} = 1.26$$

$$\sin r = \frac{\sin i}{1.26} = \frac{\sin 57}{1.26} = 0.67$$

$$r = 41.7°$$

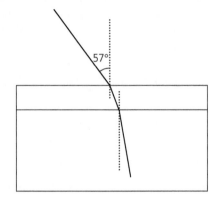

b $\dfrac{\sin i}{\sin r} = \dfrac{\sin 41.7}{\sin r} = \dfrac{1.3}{1.26}$

$\sin r = 0.64 \Rightarrow r = 40.15°$

c $i = 57°$

$$\frac{\sin i}{\sin r} = \frac{\sin 57}{\sin r} = n = 1.3$$

$\sin r = 0.64 \Rightarrow r = 40.15°$

d The answers are the same because the parallel-sided layer of oil will only laterally displace the light and make no difference to the angle at which the ray enters the water.

Total internal reflection

When light travels from a material of high refractive index to a material of lower refractive index, such as from glass to air, it bends away from the normal (Figure 18).

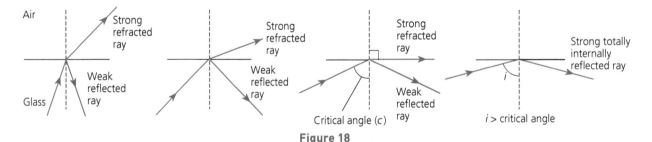

Figure 18

For small angles of incidence, most of the light will be refracted out of the glass and there will be weak reflection at the boundary. As the angle of incidence is increased, the angle of refraction will increase. When the angle of refraction is 90°, the angle of incidence is called the **critical angle**, *c*. When the angle of incidence is greater than the critical angle, all the light is reflected at the boundary and **total internal reflection** occurs.

Conditions for total internal reflection:

- Light is travelling from a material of high refractive index to a material of lower refractive index.
- The angle of incidence in the material of higher refractive index is greater than the critical angle.

Critical angle

The critical angle is therefore angle of incidence that produces an angle of refraction of 90°. The value of the critical angle depends of the refractive index of the material.

Applying Snell's law to light travelling from glass to air, we get:

$$_{glass}n_{air} = \frac{\sin i}{\sin r} = \frac{\sin c}{\sin 90} = \frac{\sin c}{1} = \sin c$$

but

$$_{glass}n_{air} = \frac{1}{_{air}n_{glass}}$$

so

$$\sin c = \frac{1}{_{air}n_{glass}}$$

or

$$_{air}n_{glass} = \frac{1}{\sin c}$$

Experiment to measure the refractive index of glass using total internal reflection and a semi-circular block

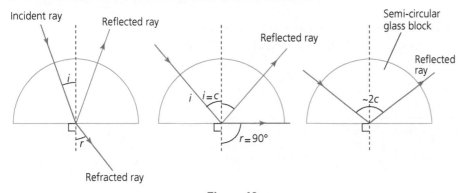

Figure 19

A semi-circular glass block is used to ensure no refraction occurs at the air–glass boundary. The light is directed through the block, towards the centre of the flat surface, so along a radius and hence along the normal to the curved surface (Figure 19). The angle of incidence *i* at the glass–air boundary is increased until →

the critical angle is reached and the angle of refraction is 90°. The light travels along the straight edge of the block. The angle of incidence may be measured at this point and is the critical angle. Alternatively, for ease of measurement, the angle of incidence could be increased by an extremely small amount so that total internal reflection takes place. The weak reflected ray becomes very bright. The angle between the incident ray and this bright reflected ray is approximately 2c. Once the critical angle is measured, the refractive index for glass can be found using:

$$_{air}n_{glass} = \frac{1}{\sin c}$$

> **Exam tip**
>
> The critical angle is not the angle of reflection when total internal reflection is occurring. The critical angle is the angle of incidence when the angle of refraction is 90°.

Applications of total internal reflection

Step index optical fibre

Optical fibres use total internal reflection to send light along a long, thin strand of glass, which has a core surrounded by a cladding of lower refractive index. Light is confined to the fibre as long as it is incident on the core–cladding boundary at an angle greater than the critical angle (Figure 20).

The optical fibre would work without the cladding as air also has a lower refractive index than the core glass (light pipe). However, the cladding is useful as it protects the core, and prevents cross talk and the leakage of light. The core should be narrow as this cuts down on multi-mode dispersion, which is where light entering the optical fibre at slightly different angles follow slightly different paths and arrives at the other end at slightly different times, causing the pulse of light to broaden out. In very high-speed systems with digital pulses travelling in quick sucession, dispersion would be an issue, and is avoided by use of a monomode optical fibre, in which the light is transmitted axially.

Optical fibres are used in instruments called endoscopes and also in communications. The use of optical fibres in communications has improved the transmission of data, giving us a more secure high-speed internet, with much greater capacity than the copper equivalent.

Many optical instruments, such as binoculars and telescopes, use 45° prisms to reflect light rather than mirrors which can cause multiple reflections and blurring (Figure 21).

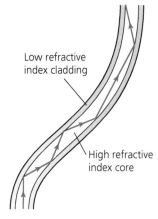

Low refractive index cladding

High refractive index core

Figure 20

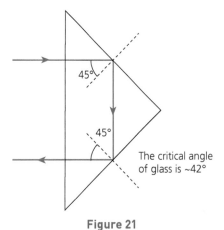

45°

45°

The critical angle of glass is ~42°

Figure 21

An optical fibre has a core of refractive index 1.48 and a cladding of refractive index 1.45. What is the critical angle for this fibre?

Answer

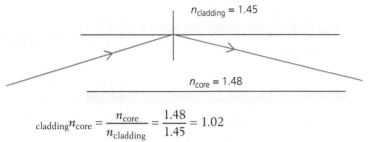

$n_{cladding} = 1.45$

$n_{core} = 1.48$

$$_{cladding}n_{core} = \frac{n_{core}}{n_{cladding}} = \frac{1.48}{1.45} = 1.02$$

and

$$n = \frac{1}{\sin c}$$

Therefore:

$$1.02 = \frac{1}{\sin c} \Rightarrow \sin c = \frac{1}{1.02} = 0.980$$

$$c = 78.4°$$

Calculate the speed of light travelling through an optical fibre made of glass with a refractive index of 1.52.

State **two** reasons why telecommunications use optical fibres rather than copper cables to transmit signals.

Flexible endoscopes

Endoscopes (Figure 22) are used to view internal parts of the body and to perform minor operations. Optical fibres are used in endoscopes.

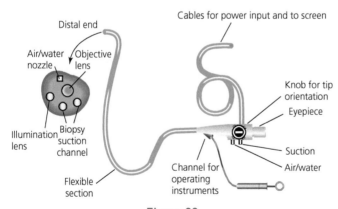

Figure 22

The flexible shaft

The shaft of the endoscope is only 10 mm in diameter and can be up to 2 m long. It is flexible, to make it easy to manoeuvre through the body, and is coated in steel and plastic in order to make it waterproof and resistant to chemical damage.

The flexible shaft includes:

- a non-coherent optic bundle — light is guided to the area under investigation by a non-coherent fibre optic bundle in which the optical fibres are randomly aligned

- a coherent optic bundle — the image must be transmitted back to the observer by a coherent fibre optic bundle of ordered, parallel fibres, which are lined up at both ends so that an image can be transmitted (Figure 23). In order to produce a clear image, the shaft contains up to 10 000 fibres.
- a distal end — this is inserted into the patient's body and is able to be bent in the desired direction. The image is focused by an objective lens on the end.
- a water pipe — water is carried to the objective lens to wash it and keep the view clear
- operations channel — tools are deployed at the distal end for surgery, for example a laser to cut tissue or seal a wound

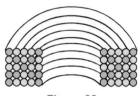

Figure 23

Uses of endoscopes

Imaging techniques are non-invasive procedures to enable medical staff to examine organs and structures inside the human body without the need for exploratory surgery. Using imaging techniques means that patients do not require a long recovery time, which they would after surgery and anaesthetic, and the risk of infection is reduced. These techniques can be deployed in outpatient departments and so avoid the cost of using hospital beds. The endoscope is just one example of a device used in such procedures.

The endoscope can be inserted into any natural opening of the body, allowing internal structures such as the oesophagus, stomach or colon to be examined. Often a small incision is made near the structure to be examined and the endoscope inserted.

- In bronchoscopy the endoscope is inserted through bronchial tubes within the lungs in order to look at the airway and to remove any objects causing a blockage.
- In gastroscopy the endoscope is inserted down the throat to look for problems with the oesophagus, stomach and duodenum, such as bleeding or ulcers.
- In laparoscopy the endoscope is inserted through an incision in the abdominal wall in order to look at abdominal organs and perform minor surgery.

The endoscope also has many industrial applications, allowing the visual inspection of inaccessible regions of complex machinery. This helps minimise the need to disassemble the machinery, saving time and money. In the aeronautical or automotive industry, for example, it might be used to look inside a valve or combustion chamber in an engine, or to search for a blockage in a climate control duct.

Knowledge check 12

Explain why a coherent bundle of fibres is used to transmit the image out of the body but non-coherent fibres are used to transmit light in to illuminate the internal structure.

Exam tip

It is important to appreciate the wide range of functions that the endoscope can carry out in addition to imaging. The tool aperture can carry a laser or a surgical implement such as a scalpel.

Summary

- At the boundary between two different media an incident wave can be reflected, transmitted (possibly refracted) or absorbed. The angle of incidence is the angle between the incident ray and the normal. The angle of reflection is the angle between the reflected ray and the normal. The angle of refraction is the angle between the transmitted ray and the normal.
- The law of reflection always demands that the angle of incidence equals the angle of reflection.

- Refraction is the change of direction of a wave caused by a change in speed as it enters a medium with different density. If the wave meets the boundary between the two materials at an angle of incidence greater than 0° (not along the normal), then the wave bends towards the normal when entering a more dense material and away from the normal when entering a less dense material. If the wave meets the boundary along the normal, there is no refraction. →

- Refraction is described by Snell's law, which states that the ratio of the sine of the angle of incidence to the sine of the angle of refraction is the same at the boundary between two different media.
- The ratio is called the refractive index and is also equal to the ratio of the speeds in the two media:

$$_1n_2 = \frac{\sin i}{\sin r} = \frac{v_1}{v_2}$$

- Total internal reflection occurs when light travels from a more dense to a less dense material and the angle of incidence is greater than the critical angle. The critical angle is the angle of incidence for which the angle of refraction is 90°. Refractive index and critical angle are related by $n = 1/\sin c$.

- The principle of total internal reflection is used in communications with fibre optic cables, medical endoscopes and reflectors using 45° prisms.
- Endoscopes use optical fibres and total internal reflection to view internal parts of the body. The flexible shaft contains a randomly arranged non-coherent fibre bundle to illuminate the area under investigation and an ordered coherent bundle of multiple parallel fibres to transmit the image back to the observer. A water channel and a tools attachment are found at the distal end. The endoscope can be inserted into any natural opening of the body. For example, it can be fed down the throat to look for problems with the oesophagus, stomach and duodenum, such as bleeding or ulcers.

Lenses

A lens is a piece of transparent material with at least one curved surface, which works by refracting light. There are two types of lens (Figure 24).

Converging lens (convex lens)

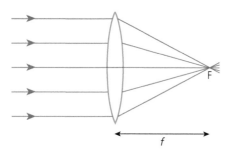

Diverging lens (concave lens)

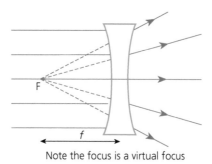

Note the focus is a virtual focus

Figure 24

Definitions

- A **convex lens** is thicker in the middle than at the edges and refracts parallel rays of light to converge to a real principal focus.
- A **concave lens** is thicker at the edges than in the middle and refracts parallel rays of light to diverge from a virtual principal focus.
- The **principal axis** of a lens is the line joining the centres of curvature of its two surfaces.
- The **principal focus**, F, of a lens is the point on the principal axis towards which all rays parallel to the principal axis converge in the case of a convex lens or from which they appear to diverge in the case of a concave lens, after refraction.
- Light can fall on either surface of a lens and so a lens has two principal foci, one on each side.
- The distance between the centre of the lens and the principal focus is the **focal length**, f, of the lens.

Exam tip

Remember that a real image is one that the rays of light actually pass through and can be formed on a screen. A virtual image is one that the rays of light only appear to pass through and cannot be formed on a screen.

Knowledge check 13

Define the principal focus of a converging lens.

Ray diagrams for a thin converging lens

It is possible to find the position and nature of the image formed by the refraction at a converging lens by constructing the predictable path taken by two of three rays (Figure 25):

- a ray parallel to the principal axis is refracted through the principal focus
- a ray through the principal focus is refracted parallel to the principal axis
- a ray through the centre of the lens is undeviated

The first two are a consequence of the definition of the principal focus, along with the reversibility of the path taken by light on refraction. The third is an approximation that applies if the lens is thin and the ray is close to the principal axis.

Note that a thin lens is represented by a straight line at which all the refraction is considered to occur; in practice it is usually refracted both on entering and leaving the lens.

(a)

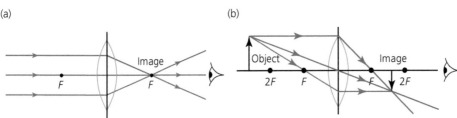

Configuration: object at infinity: point image at F

Applications: burning a hole with a magnifying glass

(b)

Configuration: object outside 2F; image between F and 2F, diminished, inverted, real

Applications: lens of a camera, human eyeball lens

(c)

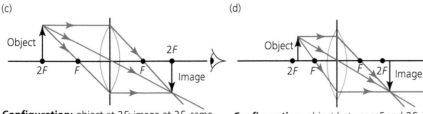

Configuration: object at 2F; image at 2F, same size as object, inverted, real

Applications: inverting lens of a field telescope

(d)

Configuration: object between F and 2F; image outside 2F, magnified, inverted, real

Applications: slide projector and objective lens in a compound microscope

(e)

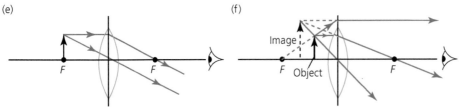

Configuration: object at F; image at infinity

Applications: lenses used in lighthouses and searchlights

(f)

Configuration: object inside F; image on the same side of the lens as the object, magnified, upright, virtual

Applications: magnifying glass, binoculars

Figure 25

> **Exam tip**
>
> Always put arrows on the rays away from the object, draw virtual rays and virtual images with dotted lines and place the eye on the opposite side from the object in line with the principal axis.

> **Knowledge check 14**
>
> Describe in full the image formed by a magnifying glass.

> **Exam tip**
>
> The use of a small, upright object, with its base on the principal axis, is a convention used to simplify matters. It ensures that the image of the base will always be on the principal axis. It must be emphasised that the rays drawn are constructional; narrow cones of rays actually enter the eye of an observer from each point of the object.

> **Exam tip**
>
> Draw ray diagrams to scale, with the values you have obtained in the experiment. This will confirm your answer and confirm that the ray diagram procedure is accurate.

Ray diagrams for a diverging lens

Figure 26 shows how to draw a ray diagram for a diverging lens.

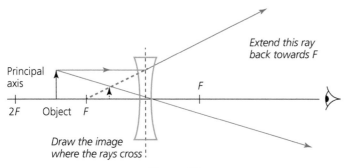

Figure 26

Lens formula

The position of an image can be calculated using the lens formula (Figure 27).

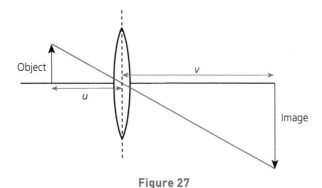

Figure 27

The lens formula is:

$$\frac{1}{f} = \frac{1}{u} + \frac{1}{v}$$

To use this formula the following are defined:

- The distance of the object from the centre of the lens is called **object distance**, u.
- The distance of the image from the centre of the lens is called the **image distance**, v.
- The distance of the principal focus from the centre of the lens is the **focal length**, f.
- The focal length of a converging lens is positive and the focal length of a diverging lens is negative.
- Distances from the centre of the lens to real images are positive and to virtual images are negative.

Exam tip

Always write the quantities u and v clearly so that the two are not confused.

Worked example

Use the lens formula to find the position an image is formed when an object is placed 20 cm from a converging lens of focal length 15 cm.

Answer

Object distance $= u = 20$ cm, image distance $= v = ?$, focal length $= f = 15$ cm.

$$\frac{1}{f} = \frac{1}{u} + \frac{1}{v}$$

$$\frac{1}{v} = \frac{1}{f} - \frac{1}{u} = \frac{1}{15} - \frac{1}{20} = \frac{1}{60}$$

$$v = 60 \text{ cm}$$

Exam tip

A common error is to forget to calculate the reciprocal of $1/v$ to find v.

Exam tip

The sign of the values is very important in the lens formula. Distances measured to real positions, objects, images or focal points are assigned positive values. Distances measured to virtual positions are assigned negative values. For example, if the image distance is positive, the image is real and is formed on the opposite side of the lens from the object.

Knowledge check 15

A virtual image is formed 8 cm from a convex lens of an object 4 cm from the lens. Calculate the focal length of the lens.

Experiment to measure the focal length of a converging lens

An *approximate* value for the focal length of a converging lens can be obtained using the light from a distant object (e.g. a window or ceiling light). The converging lens will form the image of the distant object at the principal focus. The lens is held so that a sharp image of the distant object is formed on a screen. The distance between the lens and the screen is measured and this is the focal length of the lens.

A more *accurate* method uses an illuminated object aligned with the lens and a screen on an optical bench or with a metre rule. The object is often a piece of wire mesh or cross wires (Figure 28).

Exam tip

An alternative starting procedure is to place the screen at the extreme end of the optical bench and move the lens position until a sharp image appears on the screen. If you cannot get a clear image the object distance may well be less than the focal length.

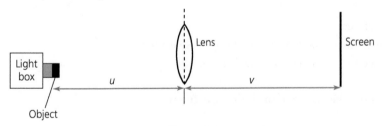

Figure 28

The object is placed at a distance greater than the approximate value obtained for the focal length.

The position of the screen is adjusted until a sharp image appears on the screen. The distance between the object and the screen, u, is measured and the distance between the image and the screen, v, is measured. The procedure is repeated for different values of object distance, u. The results are recorded in a table similar to Table 3. →

Exam tip

Alignment is very important in this experimental set-up. If no optical bench is available, a metre rule can be used to line up all the components.

Table 3 Measuring the focal length of a converging lens

Object distance, u/m	Image distance, v/m	$\dfrac{1}{u}$/m^{-1}	$\dfrac{1}{v}$/m^{-1}
0.50	0.34		
0.45	0.36		
0.40	0.41		
0.35	0.46		
0.30	0.59		

If the lens formula is rearranged and mapped to the equation of straight line we can find out which graph will give us a straight line.

$$\frac{1}{f} = \frac{1}{u} + \frac{1}{v}$$

$$\frac{1}{v} = \frac{1}{f} - \frac{1}{u}$$

$$\frac{1}{v} = -1\left(\frac{1}{u}\right) + \frac{1}{f}$$

$$y = mx + c$$

Hence a graph of $1/v$ on the y-axis and $1/u$ on the x-axis will give a straight line with a gradient of -1. The intercept on either axis is equal to $1/f$. Both intercepts are found and an average taken to determine the value of the focal length, f (Figure 29).

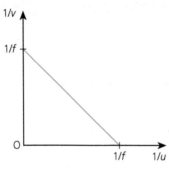

Figure 29

The linear magnification, m, of an image is the ratio of the size of the image divided by the size of the object. However, by similar triangles it can be shown that the magnification is also the object distance divided by the image distance (Figure 30).

$$\frac{AB}{AC} = \frac{DE}{CD}$$

$$\frac{O}{u} = \frac{I}{v} \Rightarrow \frac{v}{u} = \frac{I}{O}$$

So: magnification $m = \dfrac{I}{O} = \dfrac{v}{u}$

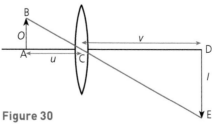

Figure 30

Knowledge check 16

A student has three lenses, one concave and two convex. He knows the concave lens has a focal length of 25 cm and the convex lenses have focal lengths of 20 cm and 10 cm, but has mixed the lenses up. State the characteristics of the lenses that he can use to identify them.

Exam tip

Graphs should use at least half of the grid, have labels and units on both axes, use an easily manageable scale, and have accurately and clearly plotted points, each marked with a small cross or a dot with a circle around it.

Knowledge check 17

Show that the intercept on both axes of the graph of $1/v$ on the y-axis and $1/u$ on the x-axis is $1/f$.

Worked example 1

An object of length 2 cm is placed 10 cm from a converging lens of focal length 15 cm. Find the length of the image.

object distance $= u = 10$ cm

focal length $= f = 15$ cm

length of object $= O = 2$ cm

Answer

First find the image distance:

$$\frac{1}{f} = \frac{1}{u} + \frac{1}{v}$$

$$\frac{1}{v} = \frac{1}{f} - \frac{1}{u} = \frac{1}{15} - \frac{1}{10} = \frac{-1}{30}$$

$$v = -30 \text{ cm}$$

$$\text{magnification} = \frac{v}{u} = \frac{I}{O}$$

$$m = (-)\frac{30}{10} = \frac{I}{2}$$

image height $= I = (-)6$ cm

This is a magnified virtual image so this lens may be a magnifying glass.

Worked example 2

Where can an object be placed in front of a converging lens of focal length 12 cm to give an image with magnification 3?

Answer

$$m = \frac{v}{u} = 3 \Rightarrow v = 3u$$

Substitute into the lens equation:

$$\frac{1}{u} + \frac{1}{v} = \frac{1}{f} \Rightarrow \frac{1}{u} + \frac{1}{3u} = \frac{1}{12}$$

$$\frac{3}{3u} + \frac{1}{3u} = \frac{1}{12} \Rightarrow \frac{4}{3u} = \frac{1}{12}$$

$$4 \times 12 = 3u \Rightarrow u = 16 \text{ cm}$$

Note that $u > f$ but $< 2f$, which is what is required to give a magnified, real image. However, this is not the only possibility — the image may have been virtual. We need to introduce the negative into our calculation:

$$m = \frac{-v}{u} = -3 \Rightarrow v = -3u$$

➜

Substitute into the lens equation:

$$\frac{1}{u} + \frac{1}{v} = \frac{1}{f} \Rightarrow \frac{1}{u} + \frac{1}{3u} = \frac{1}{12}$$

$$\frac{2}{3u} = \frac{1}{12} \Rightarrow 2 \times 12 = 3u \Rightarrow u = 8\,\text{cm}$$

Note the value of $u < f$, which is what is required to give a magnified, virtual image.

Power of a lens

The ability of a lens to focus parallel rays of light is called the power of the lens.

$$\text{power of lens} = \frac{1}{f}$$

The focal length is measured in metres and hence the power is measured in metres^{-1}. 1 metre^{-1} is equivalent to 1 dioptre $(1\,\text{m}^{-1} = 1\,\text{D})$.

A converging lens has positive power and a diverging lens has a negative power.

Worked example

A convex lens is placed at the 50 cm mark on a metre rule. Calculate the power of the lens if an image is formed at the 80 cm mark of an object that is placed at the 30 cm mark of the rule. Give your answer in dioptres.

Answer

The answer is to be in dioptres so the distances must be in metres:

$$u = 20\,\text{cm} = 0.2\,\text{m}, \, v = 30\,\text{cm} = 0.3\,\text{m}$$

$$\frac{1}{f} = \frac{1}{u} + \frac{1}{v} \Rightarrow \frac{1}{f} = \frac{1}{0.2} + \frac{1}{0.3} = 8.3\,\text{D}$$

Exam tip

When power is to be calculated, convert distances to metres before substituting into the lens formula and do not find the reciprocal at the end unless the focal length is also required.

Knowledge check 18

Compare the power of a lens with its focal length and thickness.

Defects of vision

The eye refracts light to form images on the retina at the back of the eyeball. Most of the refraction takes place at the air–cornea boundary when light slows down on entering the eye as this is where the greatest difference in refractive index occurs. This is a fixed amount of refraction, which the eye cannot change.

Refraction also takes place at the lens of the eye. The eye can change the amount of refraction by changing the shape of the lens to form sharp images of objects at different distances. It becomes thicker to give it more refracting power and a shorter focal length if it is to focus near objects. It becomes thinner to give it less refracting power and a longer focal length to focus on far objects. The ability to vary the focal length of the eye lens to focus on objects of different distances is called **accommodation**. The ciliary muscles around the lens are responsible for the change in shape of the lens:

- When they relax the vitreous humour in the eyeball pushes out to flatten the lens, making it suitable for viewing distant objects.
- When they contract, they work against the pressure of the vitreous humour and allow the lens to become fatter and suitable for viewing near objects (Figure 31).

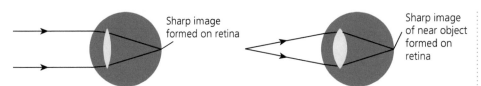

Figure 31

Note that the muscles are only working when viewing near objects, which is why we get tired when reading.

The most distant point that the unaided eye can see clearly is called the **far point**. For a normal eye the far point is assumed to be at infinity.

The closest point that the unaided eye can see clearly is called the **near point**. For a normal eye the near point is around 25 cm.

To form a clear image, the eye must focus light rays on the retina at the back of the eye. There are a number of defects that affect normal vision. The two most common defects are short sight (myopia) and long sight (hypermetropia). These can be corrected with the use of spectacle lenses.

Knowledge check 19

Explain why a swimmer finds it difficult to focus underwater.

Myopia: short sight

A person who is unable to focus distant objects suffers from the eye defect of myopia. This results in the image of a distant object appearing blurred to the viewer. They can see near objects clearly and so they are said to have short sight. Myopia is usually caused by large eyeballs.

They cannot make their lens thin enough to focus the light on their retina. It is too powerful, so the light from a point on the object converges at a point before the retina and a blurred image is experienced. A concave lens is used to correct this defect as it diverges the light prior to entry into the eye (Figure 32).

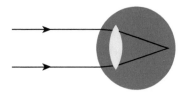

Short sight — unable to focus distant objects

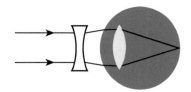

Correction method for short sight

Figure 32

Hypermetropia: long sight

A person who is unable to focus near objects suffers from the eye defect of hypermetropia. They can see far objects clearly and so they are said to have long sight. Hypermetropia can be caused by small eyeballs or weak ciliary muscles.

They cannot make their lens fat enough to focus the light on their retina. It is not powerful enough so the light converges at a point behind the retina and a blurred image is experienced. A convex lens is used to assist in the refraction and so correct this defect (Figure 33).

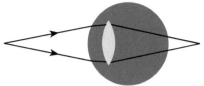
Long sight—unable to focus close objects

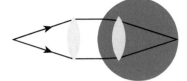
Correction method for long sight

Figure 33

Worked example

A person can focus clearly on objects between 150 cm and 400 cm.

a Define the terms near point and far point and state the near point and far point for this person.

b What power of spectacle lens would be needed to change the person's near point to a normal distance from the eye?

c Where would the person's far point be when wearing these spectacles?

d What power of lens would be required to change the person's far point to a normal far point?

e Suggest how this person might correct both these problems with one pair of spectacles.

Answer

a The near point is the closest distance at which objects can be seen clearly, unaided: 150 cm

The far point is the farthest position at which distant objects can be seen clearly, unaided: 400 cm

b $\frac{1}{f} = \frac{1}{u} + \frac{1}{v}$

A normal near point is 25 cm.

An object placed at the normal near point must form a virtual image at the actual near point of 150 cm.

$$\frac{1}{f} = \frac{1}{25} + \frac{1}{-150}$$

$$\frac{1}{f} = \frac{6}{150} - \frac{1}{150}$$

$$f = 30 \, cm$$

$$P = \frac{1}{f} = \frac{1}{0.3} = 3.33 \, D$$

c We need to find the object distance that will form a virtual image at the actual far point of 400 cm with this lens.

$$\frac{1}{f} = \frac{1}{u} + \frac{1}{v}$$

$$\frac{1}{30} = \frac{1}{u} + \frac{1}{-400}$$

$$u = 27.9 \, cm$$

Exam tip

The distance between the spectacle lens and the eye lens is ignored in the calculation.

When completing calculations involving corrective lenses, the image distance is taken as negative as the spectacle lens forms a virtual image. The object distance is where the person would like the object to be placed (the corrected near or far point) and the image distance is the actual uncorrected near point or far point of the unaided eye.

Content Guidance

So wearing the spectacles to correct the near point will give the person a far point of 28 cm.

d A normal far point is at infinity so they want to see an object at infinity and form a virtual image at the actual far point.

$$\frac{1}{f} = \frac{1}{u} + \frac{1}{v}$$

$$\frac{1}{f} = \frac{1}{\infty} + \frac{1}{-400} = -\frac{1}{400} \text{ cm}^{-1} = -\frac{1}{4} \text{ D}$$

e Use bifocal lenses.

Summary

- Lenses refract light. A converging (convex) lens converges parallel rays of light to a real principal focus.
- A diverging (concave) lens appears to diverge parallel rays of light from a virtual principal focus.
- The focal length is the distance between the centre of the lens and the principal focus. The focal length of a lens can be determined approximately using a distant object or, more accurately, using an illuminated object and a screen.
- Ray diagrams can used to graphically find the image formed by a lens. At least two rays must be used to locate the image. The paths of the three rays are possible to predict. A ray through the centre of the lens is undeviated. A ray parallel to the principal axis will be refracted through the principal focus of the lens. A ray through the principal focus will be refracted parallel to the principal axis.
- An image is fully described by its position, size (enlarged or diminished), orientation (erect or inverted) and nature (real or virtual).
- The lens formula can used to mathematically find the image formed by a lens:
$$\frac{1}{f} = \frac{1}{u} + \frac{1}{v}$$
- The magnification formula is:
$$m = \frac{\text{height of image}}{\text{height of object}} = \frac{v}{u}$$

- The power of a lens is related to the focal length:
$$\text{power} = \frac{1}{f}$$
If the focal length is measured in metres then the unit of power is the dioptre (D).
- The eye refracts light to form sharp images on the light-sensitive retina at the back of the eye. If the eye does not function perfectly, the light may not converge exactly on the retina but before or after it. This means that the image cannot be sharply focused and there is said to be a defect of vision.
- The near point is the nearest point that can be focused by the eye, normally 25 cm. The far point is the furthest point that can be focused by the eye, normally taken to be infinity.
- Myopia is short sight and the light converges before the retina. A diverging lens is used to correct this defect. The focal length is calculated using the lens formula with u taken as the corrected far point and v taken as the virtual uncorrected far point.
- Hypermetropia is long sight and the light converges behind the retina. A converging lens is used to correct this defect. The focal length is calculated using the lens formula with u taken as the corrected near point and v taken as the virtual uncorrected near point.

Knowledge check 20

What is the range of distinct vision for a person with normal sight?

Superposition, interference and diffraction

What happens when two or more waves meet? Is their motion changed, as when solid objects collide?

When waves of the same type meet, unlike particles, they pass through each other unaffected. But where they meet or cross, **superposition** will occur.

The principle of superposition states that, when two or more waves meet at a point, the resultant displacement at that point is equal to the vector sum of the individual displacements due to each wave at that point.

If two waves meet in phase, they add together to give a wave with an amplitude that is the sum of the amplitudes of the original waves. This is called **constructive interference**.

If the two waves of the same amplitude meet out of phase by 180° or π radians (antiphase), the waves cancel each other out. This is called **total destructive interference** (Figure 34).

If the amplitudes are not the same then when destructive interference takes place the resultant wave has a smaller amplitude. The two waves do not completely cancel each other out.

Exam tip

Learn the definition of superposition accurately. The principle applies to all types of wave.

Exam tip

Because displacement is a vector, we must remember to add the individual displacements, taking account of their directions.

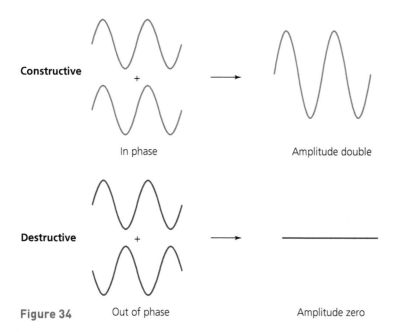

Figure 34

Graphical addition of sinusoidal waves

The principle of superposition can be used to explain many wave effects. It is applied in Figure 35, where the resultant of two waves of different amplitudes and frequencies is obtained by adding the displacements. The shape of the resultant is often very different, as in this example.

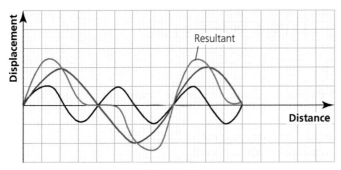

Figure 35

Exam tip

You must not say that a condition of coherence is that the waves are in phase, but rather that the waves must maintain a constant phase relationship. This is only possible if they have the same frequency. In practice coherent sources are obtained from a single source or origin.

Coherence

Wave sources are coherent if:

- the waves are the same type
- the waves have the same frequency
- the waves always maintain a constant phase difference (the phase difference might be zero but it does not have to be)

Interference patterns

When waves from two coherent sources meet at a point they superpose to produce an interference pattern. Consider the loudspeakers in Figure 36.

Figure 36

Exam tip

The light from two car headlights will overlap, but an observable interference pattern is not produced. This is because the light from the headlights does not have a constant phase difference — they are not coherent.

Non-coherent sources will still superpose, but no interference pattern is produced. The resultant conditions at each point are continuously changing.

If they are connected to the same signal generator they produce coherent sound waves, in this case in phase. There will be points along a line perpendicular to the direction of the waves where the waves meet. At a point equidistant from both sources, along the middle line, the waves have travelled the same distance and they will be in phase. Constructive interference will occur at this point and a loud sound will be detected. At other points, where one wave has travelled half a wavelength further than the other, the waves will be in antiphase and so destructive interference will occur and almost no sound will be detected at that point. This pattern of alternating maximum and minimum intensities is called an interference pattern.

Any pair of coherent sources can generate an interference pattern, including water and electromagnetic waves. For light waves the interference pattern is alternate bright and dark fringes. For water waves the interference pattern is alternate crests (or troughs) and flat water.

Interference (refraction and diffraction) of water waves can be demonstrated using a ripple tank (Figure 37a). The interference pattern for surface water waves has the significant advantage of being visible in the complete region where the coherent sources overlap (Figure 37b). This pattern exists for both sound and light.

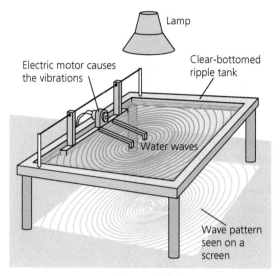

Figure 37(a)

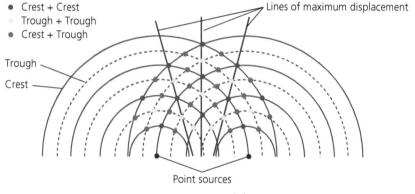

Figure 37(b)

Path difference

The difference between the distances from the two sources to a given point is called the path difference.

For two coherent sources that are in phase to start with, when the path difference is zero or equal to a whole number of wavelengths constructive interference takes place. When the path difference is an odd number of half-wavelengths, destructive interference takes place.

Knowledge check 21

State the conditions for total destructive interference.

Conditions for observable interference are as follows:

- Sources must be coherent.
- Sources should have the same amplitude so that the contrast between constructive and destructive is more evident.

Standing waves

Water waves at sea are examples of **progressive (travelling) waves**. The energy is transferred in the direction of the wave. **Standing (stationary) waves** are another example of interference; here the energy is not transferred but stored within the vibrating particles. Standing waves are set up when two waves meet that are of equal frequency and amplitude and are travelling at the same speed in opposite directions.

The graphs in Figure 38 represent two progressive waves of equal amplitude and frequency travelling in opposite directions through one another. This is normally achieved in practice by the incident wave being reflected back through itself (e.g. Melde's experiment, p. 40). The waves are viewed every quarter period. The red wave is travelling from left to right and the blue wave from right to left.

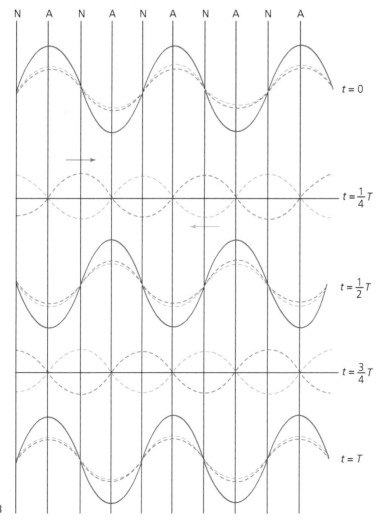

Figure 38

Initially, $t = 0$, the waves are in phase and superposition shows that the resultant purple wave has an amplitude twice that of either progressive wave. A ¼ period later, when the two progressive waves have each moved a quarter of a wavelength in opposite directions, the waves are exactly out of phase and superposition shows that the resultant is zero everywhere. Another ¼ period later, half a period from the start, the waves are again in phase, with maximum displacement for the resultant. And so the pattern continues. Finally, in the last graph, $t = T$, one period after the first, we are effectively back at the start.

The points labelled N, called nodes, are where the displacement of the resultant is always zero. The displacement of the resultant at the points labelled A, called antinodes, varies periodically from maximum to zero.

The resultant wave pattern is one of standing or stationary loops. In this example, a five-loop pattern is formed.

> **Exam tip**
>
> The wavelength of the standing resultant wave is equal to the wavelength of the progressive waves from which it is formed. This offers a solution to the otherwise difficult problem of measuring the wavelength of a progressive wave.
>
> The wavelength is twice the distance between successive nodes or successive antinodes.

> **Exam tip**
>
> Within a single loop all particles oscillate in phase but with different amplitudes.
>
> The oscillations in one loop are in antiphase with those in an adjacent loop.

Comparison of progressive and standing waves

Properties of progressive waves and standing waves are compared in Table 4.

Table 4

Progressive waves	Standing waves
Energy is transferred away from the source in the direction of wave travel	Energy is not carried away from the source but is stored within the vibrating particles
Crests and troughs move along the wave direction	Nodes and antinodes do not move along the wave direction
All points along the wave have the same amplitude	Amplitude varies between a maximum at the antinodes to zero at the nodes
Adjacent points in the wave have a different phase relationship	All points between each pair of consecutive nodes have a constant phase relationship

> **Knowledge check 22**
>
> A sound wave in an organ pipe is a standing, mechanical, longitudinal wave. Explain these terms.

> **Exam tip**
>
> At frequencies other than the natural frequencies there is no fixed pattern and the energy is dissipated quickly. When the forcing frequency of vibration equals the natural frequency, resonance occurs.

In practice, standing waves are only produced in systems with boundaries. Progressive waves reflected from the boundaries interfere with progressive waves travelling to the boundaries. Now the resulting standing waves have to comply with the additional boundary conditions, such as nodes at the ends of a stretched string or at the closed end of an air pipe, if they are to 'fit' into the system and be successfully formed. The frequencies at which these occur are called the **natural** frequencies.

Standing waves in stretched strings: transverse standing waves (Melde's experiment)

A string stretched between two fixed points can be forced to vibrate by being plucked. It will vibrate freely at all of its *natural* frequencies. If the string is forced to vibrate at a particular frequency that matches one of its *natural* frequencies, it will vibrate with a large amplitude of vibration. This is an example of resonance (Figure 39).

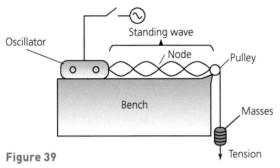

Figure 39

The signal generator makes the mechanical oscillator vibrate, causing waves to travel along the stretched string. When the waves meet the pulley they are reflected back along the string. Two coherent waves are moving in the opposite direction and a standing wave is set up if the frequency of the signal generator equals that of a natural frequency. A node always occurs at each end as the string is fixed at the ends and so the particles cannot move.

The lowest natural frequency is called the fundamental frequency, f_0, at which the string vibrates with a large amplitude in the form of a single loop (Figure 40).

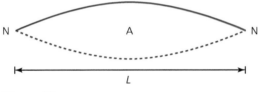

Figure 40

A single loop is half a wavelength, so the length of the string $L = \lambda_s/2$ or $\lambda_s = 2L$. But $\lambda_s = \lambda_p$. From the wave equation, $f = v/\lambda_p$, so $f_0 = v/2L$.

λ_s = wavelength of the standing wave, λ_p = wavelength of the progressive wave.

The frequency is increased and the amplitude of vibration decreases until the next frequency is reached that causes resonance and two loops are now seen. This is called the first overtone frequency, f_1 (Figure 41).

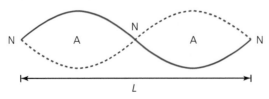

Figure 41

Two loops is a full wavelength, so $L = \lambda_s = \lambda_p$. Using the wave equation, $f_1 = v/L = 2f_0$.

As the frequency is increased further again the amplitude of vibration decreases until the next frequency is reached that causes resonance and three loops are now seen. This is called the second overtone frequency, f_2 (Figure 42).

Exam tip

The different standing wave patterns possible are referred to as the modes of vibration.

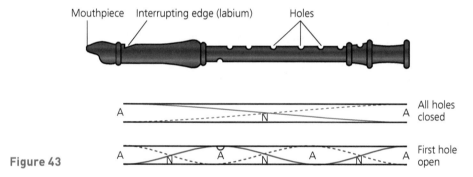

Figure 42

Three loops is one and a half wavelengths, so $L = 3\lambda_s/2$ or $\lambda_s = \lambda_p = 2L/3$. Using the wave equation, $f_2 = 3v/2L = 3f_0$.

Standing waves in air pipes closed at one end: longitudinal standing waves

Standing waves can also be set up in air pipes such as a recorder (Figure 43).

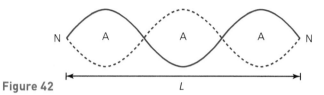

Figure 43

Standing waves in pipes can be investigated in the physics laboratory using a resonance tube — a pipe closed at one end (Figure 44).

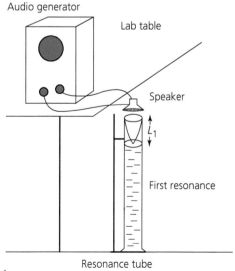

Figure 44

The frequency of the signal generator is increased from the lowest level until a clear loud sound is heard. Sound waves travel from the loudspeaker down the tube and are reflected from the end to travel back up the pipe. The waves meet and a standing wave is created at this resonant frequency. This is the fundamental frequency. The motion of the air particles is shown in Figure 45. At the open end the particle is vibrating with maximum amplitude and so this is an antinode. The vibration of the particle at the closed end is zero, so this is a node.

The displacement of the air particles is more conveniently represented by a displacement–distance graph for the vibration, although it should be remembered that the disturbances are longitudinal.

Consider the boundary conditions that apply in this case. Air cannot move at the closed end of the pipe, so this is always a node. The amplitude of vibration is a maximum at the open end, so this always an antinode. Consequently resonance will occur with a whole number of half loops or quarter wavelengths (Figure 46).

Figure 47 shows the first position of resonance when a loud, clear note can be heard. The fundamental frequency $L = \lambda/4$, so $\lambda = 4L$. Using the wave equation $v = f\lambda$ gives $f = v/\lambda$, so $f_0 = v/4L$.

Figure 48 shows the second position of resonance when a loud, clear note can be heard. The 1st overtone frequency $L = 3\lambda/4$, so $\lambda = 4L/3$. Using the wave equation $v = f\lambda$ gives $f = v/\lambda$, so $f_1 = 3v/4L = 3f_0$.

Figure 49 shows the third position of resonance when a loud, clear note can be heard. The 2nd overtone frequency $L = 5\lambda/4$ so $\lambda = 4L/5$. Using the wave equation $v = f\lambda$ gives $f = v/\lambda$, so $f_2 = 5v/4L = 5f_0$.

Figure 47

Figure 48

Figure 49

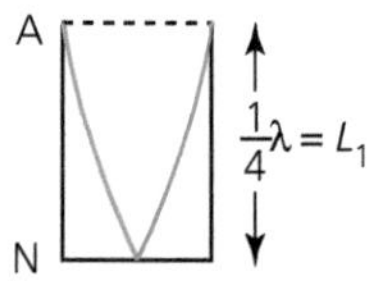

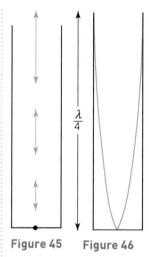

Figure 45

Figure 46

Exam tip

This time we have not shown the $\lambda_s = \lambda_p$ step, but it still applies.

Exam tip

Practise drawing the modes of vibration for standing waves in stretched string and in a pipe closed at one end and be able to label all the nodes and antinodes.

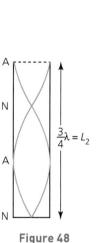

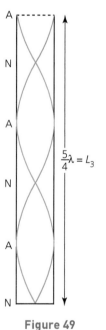

Experiment to measure the speed of sound using a resonance tube

Standing waves in a resonance tube closed at one end can be used to measure the speed of sound (Figure 50).

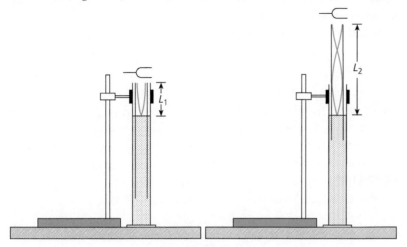

Figure 50

- A resonance tube that is open at both ends is placed in a deep tank of water. The water provides a closed end for the tube. Raising and lowering the resonance tube in the water changes its length.
- A vibrating tuning fork of known frequency is held over the end of the resonance tube in its lowest position, causing the air to vibrate.
- The tube is slowly raised, increasing the length of the air column until at a certain position the loudest note is heard. This is the first position of resonance.
- The length L_1 of the air column at this resonant position is recorded.
- The tube is raised further until a second resonance position is found. This corresponds to the second resonance position.
- The length L_2 of the air column is also noted.

At the first resonance position:

$$L_1 = \frac{\lambda}{4}$$

At the second resonance position:

$$L_2 = \frac{3\lambda}{4}$$

$$L_2 - L_1 = \frac{3\lambda}{4} - \frac{\lambda}{4} = \frac{\lambda}{2}$$

So:

$$\lambda = 2(L_2 - L_1)$$

Using the wave equation, $v = f\lambda$:

$$v = 2f(L_2 - L_1)$$

where f is the known frequency of the tuning fork in Hz, and L_1 and L_2 have been measured in m. This allows the speed of sound to be calculated in $m\,s^{-1}$.

A second method uses a range of tuning forks of different frequencies to create the standing waves for the first resonance position of each fork. This fundamental node of vibration is a quarter of a wavelength of the sound wave, so $L = \lambda/4$.

$$v = f\lambda \text{ and } \lambda = 4L$$

So:

$$v = 4fL$$

or:

$$L = \frac{v}{4f}$$

Comparing with the equation of a straight line:

$$y = mx + c$$

A graph of L on the y-axis and $1/f$ on the x-axis will be a straight line through the origin with gradient $= v/4$.

The speed of sound is approximately $340\,m\,s^{-1}$.

Young's double-slit experiment

Thomas Young demonstrated interference of light in 1801 (Figure 51). His experiment uses a monochromatic (single wavelength) light source. This source passes through a narrow single slit to create a small, well-defined source of light, which diffracts (spreads out) as it passed through the slit. The diffracted waves are then incident on a double-slit arrangement to form two coherent light sources, S_1 and S_2, separated by a distance a. The slits S_1 and S_2 are equidistant from the single slit, so the light emerges from the slits in phase. The diffracted light from the two coherent sources overlaps and interferes. An interference pattern of alternate bright and dark fringes is seen on a screen a distance d from the sources. The fringes are a distance y apart.

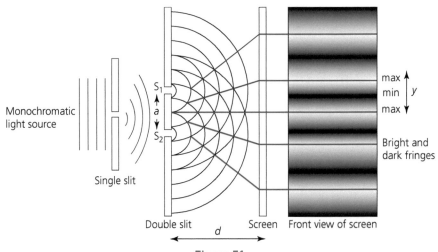

Figure 51

At a point mid-way between the two sources, the waves have travelled the same distance and are in phase, so constructive interference occurs and a bright fringe is displayed. At points either side of the mid-way point, waves will have different path lengths on arrival at the screen. The waves will give constructive interference if the path difference is a whole number of wavelengths as the waves will then be in phase and a bright fringe will be formed. The waves will give destructive interference if the path difference is an odd number of half-wavelengths as the waves will then be in antiphase and a dark fringe will occur.

This is a useful experiment, as the unknown wavelength of a source of monochromatic light can be determined by measuring the slit separation, the distance of the double slits from the screen and the fringe separation. The wavelength of the light can be calculated using the formula:

$$\lambda = \frac{ay}{d}$$

where λ = the wavelength of the light in metres, a = the separation of the double slits in metres, y = the fringe separation on the screen in metres, d = the distance between the double slit and the screen in metres.

If the double slit were illuminated by white light instead of monochromatic light, the different wavelengths making up the white light would produce its own interference fringe pattern. There would be a central white fringe with dark fringes either side. Beyond the centre the maxima and minima of the different colours would overlap and a pattern of coloured fringes produced.

Determining the wavelength of laser light using a double slit

This is, in effect, a modern equivalent of the Thomas Young experiment. A laser produces a narrow beam of coherent light. If this beam is directed at a double slit, as shown in Figure 52, then interference fringes will be visible on a screen placed some distance away —typically 2 m.

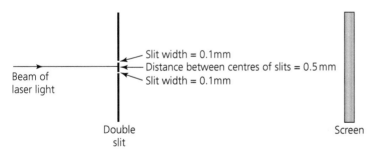

Beam of laser light

Slit width = 0.1 mm
Distance between centres of slits = 0.5 mm
Slit width = 0.1 mm

Double slit

Screen

Figure 52

The experiment is best carried out in a darkened room to improve the visibility of the fringes. A number of fringes will be visible on the screen. The distance across as many fringes as possible is measured with a rule and divided by the number of fringes in order to find the fringe separation y. Typically a manufactured double slit will have a slit separation of 0.5 mm.

Calculate the wavelength of light that produces fringes of separation 0.95 mm with coherent light from two slits 0.50 mm apart on a screen a distance 0.78 m from the slits. Give you answer in nanometres.

Answer

a = the separation of the slits = 0.50 mm = 0.50×10^{-3} m

y = the fringe separation = 0.95 mm = 0.95×10^{-3} m

d = the distance between the slits and the screen = 0.78 m

$$\lambda = \frac{ay}{d} = \frac{0.5 \times 10^{-3} \times 0.95 \times 10^{-3}}{0.78} = 6.09 \times 10^{-7} \text{ m} = 609 \text{ nm}$$

This formula is given on the formula sheet but students often mix up the different quantities. Unfortunately 'd' represents two different quantities in the two slit and grating equations. Make sure you know what each symbol represents. Always convert all the units to metres.

Diffraction

Diffraction is the spreading of waves when they pass through an opening or round an obstacle. The extent of the diffraction depends on the size of the gap compared with the wavelength. Diffraction is most noticeable if the size of the gap is approximately equal to the wavelength of the wave (Figure 53).

Describe the effect on fringe separation if the monochromatic source used in a Young's slit arrangement is replaced with one of longer wavelength.

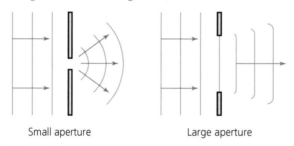

Small aperture   Large aperture

Figure 53

Diffraction explains why you can hear a noise around an open door but cannot see the source. The wavelength of the sound is about the same as the width of the door and so diffracts around it, but the wavelength of the light is much smaller than the width of the door.

Light wavelengths are very small, so diffraction effects can only be observed when the aperture (gap) is extremely small. When such a narrow slit is used with laser light, a diffraction pattern can be viewed on a screen. There is a central band of maximum intensity and a series of bright and dark bands either side (Figure 54).

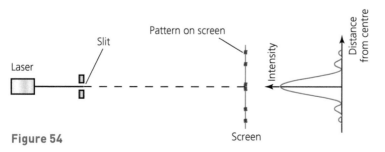

Figure 54

The wavelength of a wave does not change as a result of diffraction.

The diffraction grating

A diffraction grating is a plate on which there is a very large number of parallel, identical, very closely spaced slits. Typically, a 'fine' grating could have 300 lines per mm.

We have observed the interference pattern produced when light is incident on a double slit; if more than two slits are used, a similar effect occurs. As the number of slits is increased, the bright fringes become brighter and sharper. Although it is called a *diffraction* grating, it could be argued that it is, in fact, an interference grating (Figure 55).

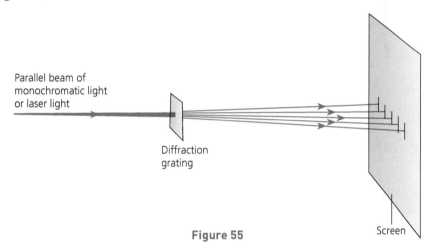

Figure 55

Suppose a beam of monochromatic light of wavelength λ falls, normally, on a grating in which the slit separation is d (Figure 56). Consider rays 1 and 2 — the path difference between these rays when they emerge at an angle θ is $d \sin \theta$. When the path difference is zero or equal to a whole number of wavelengths constructive superposition will take place. The same applies to all corresponding neighbouring rays — 2 and 3, 3 and 4, etc.

So if at the angle θ, $d \sin \theta = n\lambda$, where $n = 0$ or an integer, constructive superposition will take place and a bright fringe will be seen.

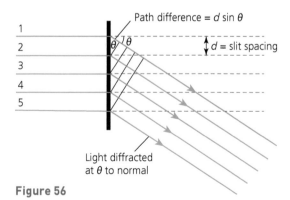

Figure 56

When $n = 0$, $\theta = 0$ and we observe in the direction of the incident light the central bright maximum, called the zero-order image, for which the path difference equals zero. First-order and second-order maxima are given by $n = 1, 2$ etc., but they are less bright than the zero-order maximum (Figure 57).

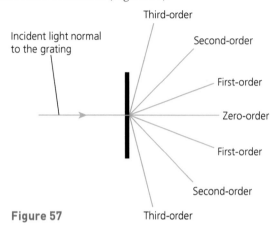

Figure 57

Exam tip

Note the symmetry either side of the central maximum.

Determining the wavelength of laser light using a diffraction grating

This is an identical set-up as for the double-slit experiment, except that a grating replaces the two slits (Figure 58).

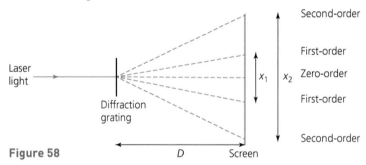

Figure 58

The wavelength of light is found using the equation for the condition for a maximum:

$$d \sin \theta = n\lambda$$

The angle θ should be calculated for more than one order and an average value for the wavelength subsequently determined.

Calculation for the first order, n = 1

The distance, D, between grating and screen is measured using a tape measure or metre rule. The distance x_1 between the first-order maxima is measured. The angle θ_1 for the first order is given by:

$$\tan \theta_1 = \frac{\frac{1}{2}x_1}{D}$$

and

$$\lambda = d \sin \theta_1$$

Worked example

Consider a grating with 500 lines per mm on which light of wavelength 600 nm falls normally.

a Find the angle at which the second-order maximum is seen.

b How many maxima in total are produced by the grating?

Answer

a Since there are 500 lines in 1 mm, $d = (1/500)$ mm.

Therefore:

$d = 2.0 \times 10^{-6}$ m

Second-order, $n = 2$

Using $d \sin \theta = n\lambda$:

$$\sin \theta_2 = \frac{2 \times (600 \times 10^{-9})}{(2.0 \times 10^{-6})} = 0.6$$

$\theta_2 = 37°$

b The maximum angle possible is 90°.

Using $d \sin \theta = n\lambda$:

$d \sin 90 = n_{max}\lambda$

$$n_{max} = (2.0 \times 10^{-6}) \times \frac{1}{(600 \times 10^{-9})} = 3.3$$

n is an integer, so $n_{max} = 3$

But the total number of maxima visible will be 7 — remember orders either side of a zero-order.

Summary

- The principle of superposition states that when waves meet the resultant displacement is the vector sum of the individual displacements due to each wave at that point.
- Constructive interference occurs when two waves that are in phase add together to give a wave with an amplitude that is the sum of the amplitudes of the original waves.
- Destructive interference occurs when two waves of the same amplitude that are in antiphase cancel each other out.
- Superposition of coherent sources causes an interference pattern that consists of alternating maximum and minimum intensities. Constructive interference occurs where waves that meet have travelled the same distance or have a path difference of a whole number of wavelengths, and arrive in phase. Destructive interference occurs where waves that meet have a path difference that is an odd number of half-wavelengths, and they arrive completely out of phase.

- Wave sources are coherent if they are the same type, have the same frequency and maintain a constant phase difference. Observable interference will occur when the waves are coherent and have the same amplitude.
- Standing waves are produced by interference of two waves travelling in opposite directions with the same frequency and amplitude.
- A node is a position on a standing wave with zero amplitude. Nodes are formed at fixed ends. An antinode is a position on a standing wave with maximum amplitude. Antinodes are formed at free ends.
- Standing waves can be formed in a stretched string. A whole number of half-wavelengths are formed in the string. The fundamental frequency is one loop, or half a wavelength, so $f_0 = v/2L$. The overtones are multiples of the fundamental frequency.
- Standing waves can be formed in air pipes closed at one end. An odd number of quarter wavelengths are formed in

→

the pipe. The fundamental frequency is one quarter of a wavelength, so $f_0 = v/4L$. The overtones are odd multiples of the fundamental frequency.

- Young's double-slit experiment demonstrates interference of light using a monochromatic light source. The double slit provides coherent sources that interfere to form bright and dark fringes on a screen. There is a central bright fringe with alternate dark and bright fringes on both sides. The wavelength of the light used can be calculated using the formula:

$\lambda = ay/d$

- Diffraction is the bending or spreading of light through an opening or round an obstacle.
- Diffraction is most noticeable if the size of the gap is approximately equal to the wavelength of the wave.
- The condition for a diffraction maximum in a diffraction grating pattern is $d\sin\theta = n\lambda$, where d is the grating spacing, θ the angle at which the maximum is observed, n is the order of the image and λ is the wavelength of the light.

Quantum physics

By the end of the nineteenth century the wave theory, despite its earlier notable successes, was unable to account for many of the observations relating to the interaction of electromagnetic radiation and matter.

Planck tried to solve the inadequacies of the wave theory. He stated that rather than electromagnetic radiation being emitted continuously, it is emitted intermittently in discrete amounts. He said that energy is quantised; it has specific values with no values in between.

A quantum of energy is a packet of energy. Energy is only considered in whole numbers of packets (quanta). Each quantum contains energy dependent on the frequency of the electromagnetic radiation.

The energy, E, of a quantum is given by:

$E = hf$

where h is Planck's constant ($= 6.63 \times 10^{-34}$ J s) and f is the frequency of the electromagnetic radiation in hertz.

From wave theory, the speed of a wave is given by $v = f\lambda$. All electromagnetic waves travel at the speed of light, c. Hence for electromagnetic waves we can write $c = f\lambda$.

So: $E = hf = \dfrac{hc}{\lambda}$

Worked example

Calculate the energy of quanta of red light of wavelength 656.3 nm.

Answer

$E = hf = \dfrac{hc}{\lambda} = \dfrac{6.63 \times 10^{-34} \times 3 \times 10^8}{656.3 \times 10^{-9}}$

$E = 3.03 \times 10^{-19} \text{J} = \dfrac{3.03 \times 10^{-9}}{1.6 \times 10^{-19} \text{eV}} = 1.89 \text{ eV}$

Exam tip

Classical physics demonstrated and explained how light waves can be reflected, refracted, diffracted, polarised and superposed. Light was well understood before the photoelectric effect was demonstrated.

The photoelectric effect

The photoelectric effect is one phenomenon that classical physics is unable to explain.

In a metal each atom has a few loosely attached outer electrons, which move randomly through the material as a whole. If one of these electrons near the surface of the metal tries to escape, it experiences an attractive inward force from the resultant positive charge left behind. An electron cannot escape the metal unless an external source does work against this attractive force and increases the kinetic energy of the electron. This can be achieved by heating (thermionic emission) or by shining electromagnetic radiation on the metal (photoelectric emission).

Photoelectric emission is the release of electrons from a metal when electromagnetic radiation of high enough frequency is incident on its surface. The electrons emitted are called photoelectrons.

The photoelectric effect can be demonstrated using a gold leaf electroscope. Ultraviolet radiation is directed onto a small sheet of zinc freshly cleaned with emery cloth to remove any oxidation, and connected to an electroscope.

When the electroscope is charged positively the ultraviolet radiation has no effect on the gold leaf.

When the electroscope is charged negatively the ultraviolet radiation causes its immediate discharge, indicated by the gold leaf returning to its vertical position (Figure 59). A sheet of glass between the ultraviolet and the zinc halts the discharge.

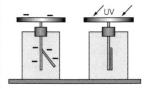

Figure 59

Laws of the photoelectric effect

- The number of photoelectrons emitted per second depends on the intensity of the incident radiation. This is justifiable in that the more intense the radiation the more energy is absorbed by the metal and the more electrons are able to escape.
- The photoelectrons are emitted with a range of kinetic energies from zero up to a maximum value, which increases with the frequency of incident radiation and is independent of intensity. This is a surprise in that classical wave theory would expect photoelectrons to have greater kinetic energy if the intensity of the radiation was greater.
- For each metal, there is a certain threshold frequency below which no photoelectric emission occurs, no matter how intense the radiation. Also, photoemission occurs immediately. There is no time delay while the electrons build up the energy required for emission. Classical wave theory cannot explain these observations.

The **work function** φ of a metal is the minimum energy that must be supplied to enable an electron to escape from its surface. Its value is characteristic of the metal type.

Worked example

The energy required to ionise an argon atom is 15.8 eV. Express this in joules.

Answer

$$1 \text{ eV} = 1.6 \times 10^{-19} \text{ J}$$

$$15.8 \text{ eV} = 15.8 \times 1.6 \times 10^{-19} = 25.3 \times 10^{-19} \text{ J}$$

Exam tip

φ is often quoted in units of electronvolts. The electronvolt (eV) is a unit of energy. It is equal to the kinetic energy gained by an electron in being accelerated by a potential difference of 1 volt. 1 eV is equivalent to 1.6×10^{-19} J. It is a much more convenient unit of energy than the joule when discussing atoms.

Einstein's photoelectric equation

Einstein extended Planck's ideas in 1905 by deriving an equation that explained the laws of photoelectric emission. He assumed that electromagnetic radiation was not only emitted in whole numbers of quanta but that they travelled in quanta and were absorbed in quanta. He called a quantum of electromagnetic energy a photon (Figure 60).

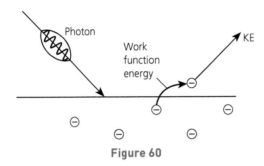

Figure 60

When a photon interacts with an electron, it transfers all its energy to a single electron instantaneously. If the frequency of incident radiation is below the threshold frequency for that metal, the energy carried by the photon is insufficient for an electron to escape. If the frequency of radiation is equal to the threshold frequency, the energy carried by each photon is just sufficient for the electrons at the surface to escape.

If the energy of the photon hf is greater than the threshold energy φ, the electron escapes and the excess energy appears as kinetic energy of the photoelectron:

$$hf - \varphi = \frac{1}{2}mv_{max}^2$$

This is Einstein's photoelectric equation.

If we apply Einstein's equation to the threshold situation, i.e. when the photon has just enough energy to liberate the electron, the electron escapes, but with no kinetic energy.

When $hf - \varphi = \frac{1}{2}mv_{max}^2 = 0$:

$$hf_0 = \varphi$$

where f_0 is the threshold frequency.

Exam tip

The electrons that escape can have any value of speed up to v_{max}. The value will depend on how much of the photon energy was actually used to escape. The remainder becomes the kinetic energy of the photoelectron. Remember that φ is the minimum energy needed to escape.

Exam tip

Note that it is assumed that a photon imparts all its energy to one electron and then no longer exists. It is also assumed that one electron will interact and receive the energy from only one photon.

Worked example

A metal surface will emit electrons if the incident radiation has a threshold wavelength of 0.65 μm. Calculate:

a its threshold frequency

b the work function in electronvolts

c the maximum speed of the electrons emitted by violet light of frequency 7.5×10^{14} Hz.

($c = 3 \times 10^8$ m s^{-1}, $h = 6.63 \times 10^{-34}$ J s, $e = 1.6 \times 10^{-19}$ C, $m_e = 9.1 \times 10^{-31}$ kg) →

Exam tip

When using all formulae, energy must be in joules not electronvolts.

Answer

a $\lambda_0 = 0.65\,\mu m = 0.65 \times 10^{-6}\,m$

$f_0 = \dfrac{c}{\lambda_0} = \dfrac{3 \times 10^8}{0.65 \times 10^{-6}} = 4.6 \times 10^{14}\,Hz$

b $\varphi = hf_0 = 6.63 \times 10^{-34} \times 4.6 \times 10^{14}\,J = \dfrac{3.05 \times 10^{-19}}{1.6 \times 10^{-19}} = 1.9\,eV$

c $hf - \varphi = \tfrac{1}{2}mv_{max}^2$

$\tfrac{1}{2}mv_{max}^2 = hf - \varphi = (6.63 \times 10^{-34} \times 7.5 \times 10^{14}) - (6.63 \times 10^{-34}) = 1.9 \times 10^{-19}\,J$

$v_{max} = \left(\dfrac{2 \times 1.9 \times 10^{-19}}{9.1 \times 10^{-31}}\right)^{\frac{1}{2}} = 6.5 \times 10^5\,m\,s^{-1}$

The Bohr model

The Bohr model of the atom, suggested in 1913, came shortly after the Rutherford nuclear model (1911). It addressed the problematic issue of electron movement in the atom and drew inspiration from the success of quantum theory and the concept of the photon.

Bohr assumed that each electron moves in a circular orbit, which is centred on the nucleus, and its energy depends on the radius of orbit. The necessary centripetal force is provided by the electrostatic attraction between the positively charged nucleus and the negatively charged electrons. He stated that electrons can only revolve around the nucleus in certain allowed orbits and while they are in these orbits they do not emit radiation.

An electron in an orbit has a definite amount of energy. It possesses kinetic energy due its motion and potential energy due to its attraction to the nucleus. Each allowed orbit is therefore associated with a certain quantity of energy that equals the total energy of an electron in it. An electron can move from one orbit of energy, E_1, to another of higher energy, E_2, by absorbing energy. An electron can fall from one orbit of energy, E_2, to another of lower energy, E_1, and the energy difference is emitted as one photon of frequency f, given by:

$\Delta E = E_2 - E_1 = hf$

(See Figure 61.)

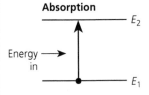

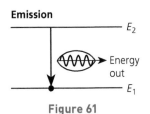

Figure 61

Energy levels

The energies of the electrons in an atom can only have certain values. These values are called the energy levels of the atom. All atoms of a given element have the same set of energy levels and these are characteristic of the element. The energies are expressed in electronvolts.

The energy levels of an atom are usually represented as a series of horizontal lines. Figure 62 is the energy level diagram of the hydrogen atom.

Hydrogen has only one electron and this usually occupies the lowest level and has energy of −13.6 eV. When the electron is in this level the atom is in its most stable unexcited state called the **ground state**.

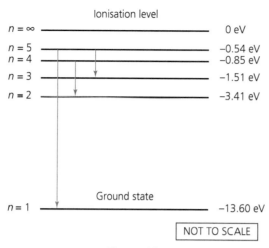

Figure 62

The atom can absorb energy by being involved in a collision with other atoms or an electron, or by absorbing heat energy or a photon of electromagnetic radiation. After absorption of energy the electron may be promoted into one of the higher energy levels (further from the nucleus). The atom is now unstable and is in an excited state. After a short time the electron will fall back into the lowest level, so the atom returns to its ground state. The energy that was originally absorbed is emitted as a photon of a specific frequency.

The emitted photon has energy equivalent to the difference in energy between the two energy levels. For example, if an electron falls from level E_2 to level E_1, then the emitted photon will have energy $E_2 - E_1$. To calculate the frequency of the emitted electromagnetic radiation the equation $E_2 - E_1 = hf$ is used.

Emission spectra

A line in an emission spectrum indicates the presence of photons of a particular frequency. It arises from the loss of energy that occurs in an excited atom when an electron falls directly or in stages to lower levels. Atoms of each element have a unique set of energy levels. Energy transitions can only occur between these specific energy levels and the electromagnetic radiation or photons emitted have energies characteristic of the element (Figure 63).

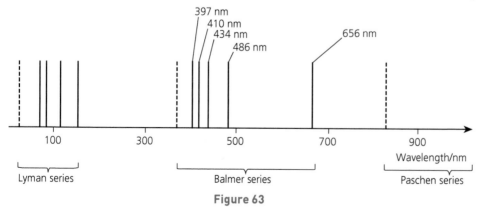

Figure 63

Exam tip

Think of an energy level diagram as a ladder of unequally spaced rungs. You require energy to climb the ladder.

Exam tip

An atom that has lost an electron is said to be ionised and therefore the energy required to ionise a hydrogen atom that is in its ground state is 13.6 eV.

Exam tip

The levels have negative values because the energy must be supplied to free an electron and ionise the atom. As a stationary free electron is considered to have zero energy, the energy levels must be negative to start with.

Exam tip

The emission spectrum is in the form of 'lines' only because it is viewed through a slit.

Worked example

With reference to the energy level diagram of the hydrogen atom (Figure 62), answer the following:

a Explain the meaning of the terms ground state and excited state.

b Write down an expression for the wavelength, λ, of the radiation emitted during a transition between two energy levels E_1 and E_2.

c Name the type of spectrum resulting from transitions between the levels in an atom, and describe its appearance.

d Find the minimum energy in joules required to remove an electron in the ground state completely from the atom.

e Identify the transition responsible for emission of light of wavelength 486 nm.

Answer

a Ground state is the lowest energy state of the electron or the atom, excited state is the state to which electrons are raised by the addition of energy.

b $E_2 - E_1 = \dfrac{hc}{\lambda} \Rightarrow \lambda = \dfrac{hc}{E_2 - E_1}$

where h is Planck's constant and c is the speed of light.

c Line spectrum — narrow, coloured lines on a dark background at specific wavelengths.

d Energy to remove electron from ground state is:

$$13.6\,\text{eV} = 13.6 \times 1.6 \times 10^{-19} = 2.18 \times 10^{-18}\,\text{J}$$

e $E_2 - E_1 = \dfrac{hc}{\lambda} = \dfrac{6.63 \times 10^{-34} \times 3 \times 10^8}{486 \times 10^{-9}} = 4.09 \times 10^{-19}\,\text{J} = 2.56\,\text{eV}$

Transition between $-0.85\,\text{eV}$ and $-3.41\,\text{eV}$

Laser action

Laser is an acronym for light amplification by stimulated emission of radiation. The action of a laser can be explained in terms of energy levels.

A material whose atoms are excited emits radiation when electrons in higher energy levels return to lower levels. Normally this occurs randomly, i.e. spontaneous emission occurs, and the radiation is emitted in all directions and is incoherent. The emission of light from an ordinary bulb is due to this process.

However, if a photon of exactly the 'correct' energy approaches an excited atom, an electron in a higher energy level may be *induced* to fall to a lower level and emit another photon. This photon will have the same frequency, phase and direction of travel as the *stimulating* photon, which is itself unaffected. (The 'correct' energy value equals the difference between the higher and lower levels.) This is called **stimulated emission** (Figure 64).

In a laser the stimulated process must be repeated time and time again. For this to happen it is necessary to have more electrons in the higher rather than lower energy level. Such a condition, called **population inversion**, is the reverse of the

Exam tip

Do not spell laser with a 'z' — the 's' stands for stimulated.

normal situation but is essential if light amplification is to occur. It is also important that the higher energy level is a **metastable level**, i.e. electrons can remain in this level for approximately 1 millisecond as compared with 10 nanoseconds in other excited levels. One method of creating a population inversion is known as optical pumping.

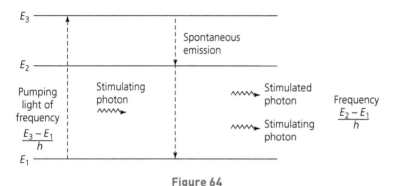

Figure 64

Uses of lasers

Lasers have a wide range of practical applications. Some examples are listed below:
- keyhole surgery and eye surgery — lasers make precision instruments
- fibre-optic communication — lasers transfer huge amounts of information at the fastest speeds
- barcode scanning — products have unique identity numbers read quickly by lasers
- reading DVDs and CDs — lasers retrieve data by reflecting off microscopic pits on the discs

Characteristics of laser light

- Monochromatic — photons have the same energy and hence frequency.
- Coherent light — photons are all in phase.
- Intense light — photons are:
  - in phase, so constructive superposition gives high amplitude
  - collimated, as all the photons travel in the same direction

Exam tip

Note the difference between the definition of 'coherence' in this context and that for interference.

X-rays in medicine: CT scanning

X-rays can be used to examine internal structures in the body because the various types of body tissue absorb different amounts of X-ray radiation. X-rays penetrate soft tissue but are stopped by bone. Photographic film is sensitive to X-ray exposure and a shadow image of the bone is obtained.

Computed tomography (CT) imaging, also known as 'CAT' (computed axial tomography) scanning, uses X-ray technology to create detailed three-dimensional images.

X-rays are a form of ionising electromagnetic radiation and can therefore cause damage to living cells (especially to the DNA) as some of the energy of the X-rays will be absorbed by the body tissue. X-rays have very high frequencies and very short wavelengths. Their wavelengths range between 0.001 nm and 10 nm.

The X-ray tube

Electrons are emitted by thermionic emission from the heated, negatively charged filament and accelerated towards the positively charged tungsten target (Figure 65).

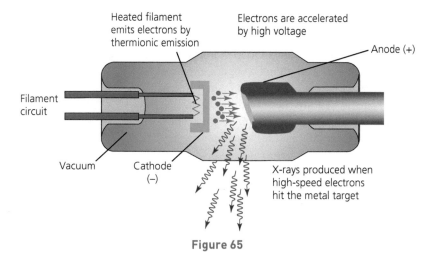

Figure 65

The X-ray tube is evacuated to ensure the electrons do not collide with any gas particles and deviate from their path. The tungsten target absorbs the electrons and releases some of the energy in the form of X-rays by one of two methods:

- the rapid deceleration of electrons as they pass near the nucleus (Bremsstrahlung)
- the tightly bound inner electrons being knocked out of atoms by the incident electrons and higher-state electrons dropping down to fill the vacancy

This process is inefficient because there is a large amount of energy released as heat. For this reason the tungsten target has a copper mounting to conduct heat and it is cooled by circulating oil through the mount. Spinning the tungsten target at high speed also helps to stop it overheating. Narrower beams of X-rays produce a sharper image. The tungsten target is therefore angled so that a wide beam of electrons produces a narrow beam of X-rays

Using X-rays to form images

- Hard X-rays are X-rays with a higher frequency and are more penetrating than soft X-rays. Soft X-rays are usually filtered out when doing a scan because they cannot penetrate through a patient's body and add needless risk of radiation damage.
- Attenuation is a measure of how much something absorbs X-rays. The amount of attenuation increases with the number of protons in the nuclei. For example, bones have a higher attenuation than soft tissue and therefore produce a dark shadow when X-rayed, whereas soft tissue appears much fainter.
- When patients have an X-ray, they are usually scanned at a frequency of approximately 7×10^8 Hz because body tissues absorb this frequency the best.
- X-rays are best suited to imaging bones and have a very high resolution. For imaging soft tissue, however, there is very little contrast and so a contrast medium is needed. Contrast mediums are substances given to the patient that absorb X-rays and produce an image of the area under investigation when X-rayed.

Exam tip

Only the production of X-rays by the process of movement between energy levels will be examined. Whilst the energy of a visible light photon is a few electronvolts, that of an X-ray is several kiloelectronvolts, indicating that we are dealing with electrons in the lower levels, held strongly by the nucleus.

Planar, conventional X-ray images are always two-dimensional projections of three-dimensional objects. In addition, there is always the possible problem of one object hiding another. One way to overcome these problems is to image successive two-dimensional slices of the patient to build up a three-dimensional image. This is called tomography, from the Greek word *tomos*, meaning a section.

Computed tomography (CT) scans are better than conventional X-rays for imaging soft tissue. In fact, CT has the unique ability to image a combination of bone, soft tissue and blood vessels. CT requires the use of powerful digital computing to process multiple images of slices of the body taken by a rotating X-ray device from different angles. A number of detectors around the patient capture the transmitted X-ray images and the three-dimensional image is built up from the combined data. The tube voltages and exposures are greater than for conventional X-radiography.

Knowledge check 25

Explain how X-rays are produced when accelerated electrons hit a metal target.

Comparison of CT with conventional X-rays

- CT images are three-dimensional and therefore more detailed.
- CT can distinguish similar tissues types, but conventional X-ray can only distinguish bone and soft tissue.
- CT involves a much greater dose of ionising radiation.
- CT requires complex computational capability.
- In CT scans the X-ray tube moves and it is prone to breaking.

Summary

- Planck stated that energy is quantised in discrete, whole packets. The energy, E, of a quantum is given by $E = hf$, where h is Planck's constant $(6.63 \times 10^{-34}\,\text{J s})$ and f is the frequency of the electromagnetic radiation in hertz. A quantum of electromagnetic energy is called a photon.
- Electromagnetic radiation can cause electrons to be emitted from a metal surface. This is called the photoelectric effect. The electrons that are released are called photoelectrons.
- Photoemission will only occur when the frequency of incident radiation is greater than the threshold frequency. Different metals have different threshold frequencies. The rate of photoemission depends on the intensity of radiation. Photoemission happens immediately. The maximum kinetic energy of the photoelectrons is dependent on the incident frequency.
- The results of the photoelectric effect conflicted with classical wave theory. It was evidence for the particle-like nature of light, as it can only be explained if electromagnetic radiation hits the metal in packets of energy.

- Einstein's photoelectric equation is:
    $$hf - \Phi = \tfrac{1}{2}mv_{max}^2$$
- The minimum amount of energy required for an electron to escape the surface is the work function $\Phi = hf_0$, where f_0 is the threshold frequency.
- The electronvolt (eV) is a unit of energy. 1 eV is equivalent to $1.6 \times 10^{-19}\,\text{J}$.
- The energies of the electrons in an atom can only have certain values, called energy levels. All atoms of a given element have the same set of energy levels and these are characteristic of the element.
- The energy levels of an atom are usually represented as a series of horizontal lines.
- The lowest energy state of an atom is its ground state. An atom can absorb energy and promote an electron to a higher energy level. The atom is now unstable and is in an excited state.
- The electron will fall back to a lower energy level with the emission of a photon of energy: $E_2 - E_1 = hf$. The photons form an emission line

- spectrum with bright lines at specific frequencies on a dark background.
- The levels have negative values because the energy must be supplied to free an electron and ionise the atom.
- Laser is an acronym for light amplification by stimulated emission of radiation.
- An electron in an excited state can be stimulated to fall to a lower energy level. A stimulating photon of exactly the right frequency can induce this fall, and a photon of the same frequency, in phase and travelling in the same direction is emitted. These two identical photons can then stimulate more identical photons to be emitted and the chain reaction continues.
- Laser light is monochromatic, coherent, intense and collimated.
- Lasers can be used in surgery, fibre optic communication, barcode scanning and reading DVDs and CDs.

- Computed tomography (CT) scans use X-rays to image the body.
- X-rays penetrate soft tissue but are stopped by bone. Photographic film is sensitive to X-ray exposure and a shadow image of the bone is obtained.
- X-rays are ionising radiation so the body tissue absorbs some of the energy of the X-rays and the living cells are damaged.
- An X-ray tube produces X-rays by bombarding a metal target with accelerated electrons.
- Conventional X-rays are used to produce planar 2-dimensional images.
- CT scans use a moving X-ray tube and a range of detectors to produce slices of the body, which are computed to form a three-dimensional image. The body receives more ionising radiation in a CT scan than in a conventional X-ray.

Wave–particle duality

Many phenomena are fully explained by the wave theory of light. For example, diffraction, interference and polarisation are evidence that the wave nature of light is a valid theory. However, the photoelectric effect requires another explanation of the behaviour of light. Planck and Einstein declared that light and other forms of electromagnetic radiation were emitted and absorbed as quanta, called photons. This implies that electromagnetic radiation could display particle-like behaviour when being emitted or absorbed. So light has an apparent dual nature — on occasions it behaves as if a wave and in other situations as if it is particles (Table 5).

Table 5

Phenomenon	Can it be explained using the wave model?	Can it be explained using the photon model?
Reflection	Yes	Yes*
Refraction	Yes	No
Polarisation	Yes	No
Interference	Yes	No
Diffraction	Yes	No
Photoelectric effect	No	Yes
Emission spectra	No	Yes
Laser	No	Yes

* Reflection can be explained if light is treated as particulate in nature.

What of the reverse situation? If a 'wave' can behave as a 'particle', could it be that what was up to now thought of as particle, behaves as a wave?

In 1923 De Broglie proposed that matter also exhibits wave–particle duality. He suggested that a particle with momentum p has an associated wavelength λ given by the formula:

$$\lambda = \frac{h}{p}$$

where λ is the de Broglie wavelength in m, p is the momentum of the particle in $\mathrm{kg\,m\,s^{-1}}$ and h is Planck's constant.

Worked example

What is the de Broglie wavelength of an electron accelerated through a potential of 1000 V?

Answer

The kinetic energy gained by the electron will be equal to the charge on the electron multiplied by the potential.

$$\tfrac{1}{2}mv^2 = eV = 1.6 \times 10^{-19} \times 1000 = 1.6 \times 10^{-16}$$

So

$$v = \left(\frac{2 \times 1.6 \times 10^{-16}}{9.1 \times 10^{-31}}\right)^{\frac{1}{2}} = 1.9 \times 10^7\,\mathrm{m\,s^{-1}}$$

momentum $p = mv$

So

$$\lambda = \frac{h}{p} = \frac{h}{9.1 \times 10^{-31} \times 1.9 \times 10^7} = 3.9 \times 10^{-11}\,\mathrm{m}$$

Electron diffraction

The electron wavelength calculated in the above worked example is comparable to the wavelength of X-rays. X-rays can be diffracted as they are waves, so electron diffraction would confirm particles exhibiting wave-like behaviour. For observable diffraction the space through which the wave passes must be of the order of the wavelength of the wave. X-rays are diffracted by the spacing between crystals, so electrons must be diffracted by the same structures.

The wave nature of particles was confirmed by electron diffraction (Figure 66). A crystal of graphite was used to diffract the accelerated electrons and an interference pattern was produced very similar to the concentric rings produced by X-ray diffraction. If the accelerating voltage was increased the spacing of the rings decreased, showing that the electrons had a shorter wavelength.

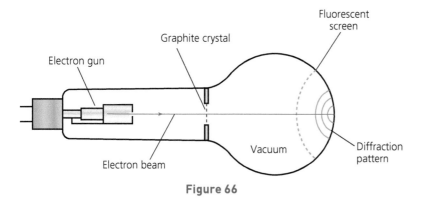

Figure 66

Worked example

Calculate the de Broglie wavelength of a 60 g ball travelling at $40\,\mathrm{m\,s^{-1}}$. Comment on the magnitude of this wavelength.

Answer

$$\lambda = \frac{h}{p} = \frac{6.63 \times 10^{-34}}{0.06 \times 40} = 2.8 \times 10^{-34}\,\mathrm{m}$$

Compared with the mass of an electron, this ball is massive and therefore its wavelength is very small to produce an observable wave-like effect. Hence objects like this and larger are characterised by their particle nature.

Summary

- Wave experiments such as Young's slits, diffraction or polarisation demonstrate the wave nature of light. The photoelectric effect demonstrates the particle nature of light. This dual nature is called wave–particle duality.
- Electrons exhibit wave–particle duality. Electrons behave as particles when they collide.
- Electrons also behave as waves as they can be diffracted by a thin sheet of graphite. For observable diffraction the wavelength should be approximately equal to the size of the gap. The atomic spacing of the graphite is used as the gap for electron diffraction. The resulting diffraction pattern is a series of concentric circles. The radius of the circles increases with increasing speed of the electrons.
- De Broglie stated that the wavelength of a moving particle with momentum p is given by $\lambda = h/p$.

■Astronomy

The expanding universe

When Edwin Hubble analysed the light from stars in distant galaxies he observed an increase in the wavelength of the characteristic lines of the absorption spectra of known elements. He concluded that this was a consequence of the galaxy moving away from us, in other words the universe is expanding.

The displacement of the spectral lines towards the red end of the visible spectrum, the red shift, can be explained as a phenomenon similar to the Doppler effect.

Doppler effect

We have all noticed at some time that the pitch (or frequency) of the siren of a police car, ambulance or fire engine appears to drop suddenly as it passes. This apparent change in the frequency when there is relative motion between source and observer is called the Doppler effect (Figure 67). It occurs with electromagnetic waves as well as sound, as illustrated by the use of the microwave Doppler effect in police radar speed checks.

 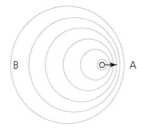

Source of sound at rest   Source of sound moving to the right

Figure 67

Consider the case of a source of sound moving to the right with velocity v, as in Figure 67. The source emits waves of frequency f with the velocity of sound c. An observer at A will hear a sound of higher frequency (lower wavelength) because the source is moving in the direction of A and 'catching up' with its emission. The net effect is that the waves are bunched together.

The decrease in wavelength to a value λ_{obs} is given by the equation:

$$\lambda_{obs} = \frac{c - v}{f}$$

For the observer at B, the opposite effect — a sound of lower frequency (higher wavelength) — is heard. As the source moves away whilst emitting waves it has the effect of stretching the waves.

The increase in wavelength observed to a value λ_{obs} is given by the equation:

$$\lambda_{obs} = \frac{c + v}{f}$$

A similar outcome occurs with light. If we observe the light to have a shorter, decreased wavelength than expected — **blue shifted** — it is because the relative motion of the source is towards us. If the light observed has a higher, increased wavelength — **red shifted** — the relative motion of the source is away from us.

Red shift

The positions of certain wavelengths (spectral lines) due to an identifiable element in the star's spectrum are compared with their positions in a spectrum of the element produced in a laboratory. **Red shift** shows the recession of the star away from the Earth. We can use the value of the shift to calculate the recession speed:

$$\lambda_{obs} = \frac{(c + v)}{f}$$

or:

$$\lambda_{obs} = \frac{(c + v)\lambda_{res}}{c}$$

> **Exam tip**
>
> The absorption spectrum of an element is the same as its emission spectrum except that it consists of dark lines on a bright background, whilst the emission spectrum is made up of bright lines on a dark background. The absorption spectrum occurs when white light passes through a cool gas or vapour. The lines are characteristic of the elements present in the vapour.

> **Exam tip**
>
> If a source of sound is approaching an observer with constant velocity, the apparent frequency of the note does not increase as the source gets nearer, it is at a constant higher value. Then as it moves past at a constant velocity a note of constant lower frequency is heard. The change occurs suddenly as the source passes.

Rearranged:

$$\lambda_{obs} = \lambda_{rest}(1 + \frac{v}{c})$$

or:

$$\frac{\lambda_{obs} - \lambda_{rest}}{\lambda_{rest}} = \frac{v}{c} = \frac{\Delta\lambda}{\lambda}$$

Most galaxies, especially distant ones, are moving away from us. This is because the universe is expanding, following the Big Bang. $\Delta\lambda/\lambda$ is called the **red shift parameter**, z, for these galaxies.

The red shift parameter can be used to find the recession speed as long as it is moving much slower than the speed of light, which is not always the case. For cases where the recession speed approaches the speed of light, for example quasars, the Doppler equations must be modified using the special theory of relativity.

For sources moving with small speeds (typically, less than half the speed of light):

$$z = \frac{\Delta\lambda}{\lambda} = \frac{\Delta f}{f} = \frac{v}{c}$$

Note: v should be treated as the relative speed between source and observer.

The red shift parameter, z, is defined as the ratio of the change in wavelength to that of a stationary source.

Figure 68 compares some of the spectral lines of hydrogen from the Sun ($z = 0$), with those for galaxies with red shifts of 0.05 and 0.10. They show that the relative arrangement is unchanged, but it is moved towards the red end of the spectrum.

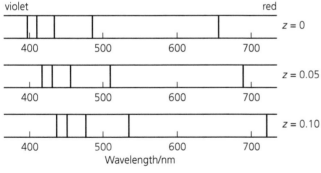

Figure 68

Worked example

The spectrum of light from the galaxy Virgo has a prominent dark line at a wavelength of 724 nm. The same spectral line observed in a laboratory on Earth has a wavelength 696 nm. How fast is Virgo moving relative to Earth?

Answer

wavelength shift, $\Delta\lambda = 724 - 696 = 28$ nm

$$z = \frac{\Delta\lambda}{\lambda} = \frac{v}{c}$$

$\rightarrow$

Therefore:

$$z = \frac{28}{696} = \frac{v}{(3.00 \times 10^8)}$$

And:

$$v = 3.00 \times 10^8 \times \left(\frac{28}{696}\right) = 1.21 \times 10^7 \, \text{m s}^{-1}$$

The wavelength has increased — z is positive — so Virgo is moving away from the Earth.

Cosmological red shift

Although cosmological red shift appears to be a similar effect to the more familiar Doppler shift, there is a subtle distinction. In Doppler shift, the wavelength of the emitted radiation depends on the motion of the body as it is releasing photons and the motion of the observer as it is received. Cosmological red shift results from the expansion of space itself and *not* from the motion of individual bodies. The wavelength is lengthened as it travels through the expanding space.

These two effects cannot be distinguished from one another by observing the spectrum and the same equations can be applied to both. The more distant the object the greater will be the impact of the expanding space through which the light moves. Both phenomena are characteristic of an expanding universe.

Hubble's law

Hubble's data analysis from a large number of galaxies showed that the recession speed v of a galaxy is proportional to its distance d from the Earth (the point of observation). Hubble's law is written as:

$$v = H_0 d$$

where H_0 is the Hubble constant, which represents the rate at which the universe is expanding, at the present time. The value of the Hubble constant is not quoted exactly owing to the difficulty in determining the distances of galaxies that are hundreds of millions of light years away.

The typical values quoted for the Hubble constant are: $H_0 \approx 2.4 \times 10^{-18} \, \text{s}^{-1}$ or $60\text{–}80 \, \text{km s}^{-1} \, \text{Mpc}^{-1}$.

Knowledge check 26

Show that 1 ly is equivalent to 9.46×10^{15} m.

Exam tip

Stars and galaxies are so far away that new units of measurement have evolved:

- the light year (ly), equal to the distance light will travel in 1 year; 1 ly = 9.46×10^{15} m
- the parsec (pc), equal to 3.08×10^{16} m or 3.26 ly
- the astronomical unit (AU), defined as the mean Earth–Sun distance, equal to 1.50×10^{11} m

Knowledge check 27

Show that the Hubble constant values 2.4×10^{-18} s^{-1} and 74 km s^{-1} Mpc^{-1} are equivalent.

The age of the universe

For a galaxy at the edge of our universe, receding from the start of time with a velocity v, Hubble's law gives:

$$v = H_0 d$$

But:

$$v = d/t$$

Therefore:

$$H_0 = 1/t$$

where t is the time elapsed from the Big Bang.

Therefore:

$$t \approx (\, 1 / 2.4 \times 10^{-18}) \approx 4.2 \times 10^{17}\,\text{s}$$

$$\approx 1.3 \times 10^{10}\ \text{years}$$

This is only an estimate because, as well as not being known accurately, the recession speed of galaxies and hence the Hubble constant are thought to have changed with time. However, the value obtained is in agreement with values obtained from radioactivity and stellar evolution calculations.

Summary

- The universe is expanding, but the rate of expansion is continually decreasing. This is because, as the galaxies move outwards, their potential energy increases and consequently their kinetic energy decreases.
- Waves change in wavelength and in frequency when given out by a moving object. The Doppler equation relates the change in wavelength $\Delta\lambda$ to the object's relative speed v:

$$\frac{\Delta\lambda}{\lambda} = \frac{v}{c}$$

- The light is red shifted for retreating objects and blue shifted for approaching objects.
- Distant galaxies are red shifted due to the expansion of the universe
- The ratio $\Delta\lambda/\lambda$ is called the red shift parameter, z, of a galaxy.
- Hubble's law, recessional speed $v = H_0 d$, is used to estimate the distance d to a distant galaxy.
- The age of the universe t can be estimated using $t = 1/H_0$.

Questions & Answers

The unit assessment

Unit AS 2 is a written examination of duration 1 hour 45 minutes. It consists of a number of compulsory, short, structured questions. Some of the questions may require an extended response of several sentences. The examination is designed to assess your understanding of all elements in the specification for this unit and all questions must be attempted. It is therefore essential that you revise all sections of the unit.

You must ensure your responses are legible and that spelling, punctuation and grammar are accurate. Use well-structured sentences starting with a capital letter and ending with a full stop. Also, present information clearly, in a logical sequence, and use appropriate scientific language.

Some examination questions will require you to demonstrate your knowledge and understanding of physics, and some questions will require you to apply this understanding to unfamiliar situations. It is important to remember that when presented with an unfamiliar situation, the principles of physics are the same and you have all the tools at you disposal to solve the problem. Be confident in your approach to the questions.

Command terms

Examiners use certain words that indicate the type of response required. It is helpful to be familiar with these terms.

- **State** — an exact, concise statement in words.
- **Define** — a word equation or a symbol equation with all the terms defined.
- **Explain** — an extended answer using correct physics terminology. The depth of the answer should reflect the number of marks available.
- **Describe** —a fuller answer, which will often be enhanced with an appropriate diagram.
- **Calculate** — a numerical answer, showing all working out. The number of significant figures in the answer should be consistent with the data, but each stage of the calculation should be kept in full in your calculator to avoid excessive rounding.
- **Determine** — the value cannot be obtained directly but some data may be extracted from another source, e.g. a graph, and used to obtain the answer.
- **Show** — a value is given and you must perform a calculation, showing all your working to lead to this value. The value should not be used in the calculation. Give your answer to more significant figures than the given value to prove you have done the calculation.
- **Sketch** — usually a graph showing a specific trend, with the axes labelled, including the origin if appropriate, and a scale if numerical data are given.

- **List** — a series of words or terms, possibly in a specific order.
- **Draw** — a carefully drawn diagram, which is fully labelled and includes all available measurements.
- **Estimate** — a calculation involving a reasonable assumption of one of the quantities used, leading to an answer of a certain order of magnitude.

Remember the following points:

- State definitions accurately.
- Always write down the formula you are using in a calculation.
- Show all substitutions and working out.
- Check you have included the correct units.
- Use the correct number of significant figures.
- Use a ruler and pencil to draw simple diagrams accurately and neatly.
- Label diagrams fully.
- Know how to use your calculator — you will need it.

Mathematics useful in Unit AS 2

- Standard form is a convenient way of writing down a very large or a very small number. The number is written as a number between 1 and 10 multiplied by 10 to the appropriate power — for example, quoted to 3 s.f. $345\,000 = 3.45 \times 10^5$.

  Be familiar with the prefixes that represent decimal multiples and submultiples of units.

Prefix	Symbol	Multiplying factor
kilo	k	10^3
mega	M	10^6
giga	G	10^9
tera	T	10^{12}

Prefix	Symbol	Multiplying factor
centi	c	10^{-2}
milli	m	10^{-3}
micro	μ	10^{-6}
nano	n	10^{-9}

- Round numbers to the appropriate number of significant figures.
- All non-zero digits are significant.
- Zeros between non-zero digits are significant, for example 5.06 (3 s.f.).
- Leading zeros are never significant, for example 0.0506 (3 s.f.).
- In a number with a decimal point, trailing zeros — those to the right of the last non-zero digit — are significant, for example 5.060 (4 s.f.).
- In a number without a decimal point, trailing zeros may or may not be significant. Explicit information on errors is needed to clarify the significance of trailing zeros, for example $74\,500$ could be to 3, 4 or 5 s.f. It is best to quote a number in standard form, i.e. a number between 1 and 10 times a power of 10, to best see the number of significant figures.
- Degrees and radians:
  - The angle between two lines can be measured in degrees or radians. In a full circle (360°) there are 2π radians. Angles can be converted between degrees and radians as follows:

    $$\text{angle in radians} = \frac{2\pi}{360} \times \text{angle in degrees}$$

- Sine and cosine graphs in terms of degrees and radians:

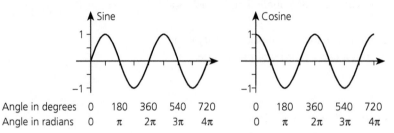

| Angle in degrees | 0 | 180 | 360 | 540 | 720 | | 0 | 180 | 360 | 540 | 720 |
| Angle in radians | 0 | π | 2π | 3π | 4π | | 0 | π | 2π | 3π | 4π |

There will be a *Data and Formulae Sheet* inside the Unit AS 2 examination paper. The constants given are as follows:

- speed of light in a vacuum $\quad\quad\quad c = 3.00 \times 10^8\,\mathrm{m\,s^{-1}}$
- elementary charge $\quad\quad\quad\quad\quad\quad e = 1.60 \times 10^{-19}\,\mathrm{C}$
- the Planck constant $\quad\quad\quad\quad\quad h = 6.63 \times 10^{-34}\,\mathrm{J\,s}$
- mass of electron $\quad\quad\quad\quad\quad\quad m_e = 9.11 \times 10^{-31}\,\mathrm{kg}$
- mass of proton $\quad\quad\quad\quad\quad\quad\quad m_p = 1.67 \times 10^{-27}\,\mathrm{kg}$
- acceleration of free fall on the Earth's surface $\quad g = 9.81\,\mathrm{m\,s^{-2}}$
- electron volt $\quad\quad\quad\quad\quad\quad\quad\quad 1\,\mathrm{eV} = 1.60 \times 10^{-19}\,\mathrm{J}$
- Hubble constant $\quad\quad\quad\quad\quad\quad\quad H_0 = 2.4 \times 10^{-18}\,\mathrm{s^{-1}}$

The formulae given that are relevant to Unit AS 2 are as follows:

- diffraction grating $\quad\quad\quad\quad\quad\quad d \sin \theta = n\lambda$

- two-source interference $\quad\quad\quad\quad \lambda = \dfrac{ay}{d}$

- lens formula $\quad\quad\quad\quad\quad\quad\quad \dfrac{1}{u} + \dfrac{1}{v} = \dfrac{1}{f}$

- Einstein's photoelectric equation $\quad \tfrac{1}{2}mv_{max}^2 = hf - hf_0$

- de Broglie formula $\quad\quad\quad\quad\quad \lambda = \dfrac{h}{p}$

- red shift $\quad\quad\quad\quad\quad\quad\quad\quad z = \dfrac{\Delta\lambda}{\lambda}$

- recession speed $\quad\quad\quad\quad\quad\quad z = \dfrac{v}{c}$

- Hubble's law $\quad\quad\quad\quad\quad\quad\quad v = H_0 d$

About this section

This section consists of two self-assessment tests. Try the questions without looking at the answers, allowing 1 hour 45 minutes for each test. Then check your responses against the answers and the comments to find out how you might improve your performance.

For question parts worth multiple marks, ticks (✓) are included in the answers to indicate where the examiner has awarded marks.

Comments on some questions are preceded by the icon ⓔ. They offer tips on what you need to do to gain full marks. Some student responses are followed by comments, indicated by the icon ⓔ, which highlight where credit is due or could be missed.

■Self-assessment test 1

Question 1

Figure 1 shows a progressive wave travelling at a speed of $12\,cm\,s^{-1}$.

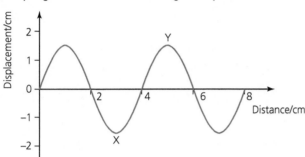

Figure 1

(a) State the amplitude of the wave. (1 mark)

(b) Determine the wavelength of the wave. (2 marks)

(c) What is the phase difference between points X and Y on the wave? (1 mark)

(d) (i) Define the periodic time of a wave. (1 mark)

(ii) Calculate the periodic time of the wave. (3 marks)

Total: 8 marks

ⓔ If a question includes a graph you must first check the label on each axis. This is a displacement–distance graph. Do not confuse this with a displacement–time graph. The wavelength can be obtained directly from a displacement–distance graph by determining the distance between successive crests. Periodic time cannot be obtained directly from a displacement–distance graph.

(a) 1.5 cm ✓

ⓔ The amplitude is the maximum displacement from the mid-point of the oscillation, so in this graph it is the displacement from the horizontal axis to the top of a crest. It is not from a crest to a trough as this is twice the amplitude. Read the scale carefully.

(b) 4 ✓ cm ✓

ⓔ The wavelength can be taken directly from a displacement–distance graph as the distance between two crests. Check that the units are correct.

(c) π radians ✓

(e) In considering phase difference in a displacement–distance graph we are comparing the motion of different particles at a single instant along a section of a wave. A particle at X is out of phase with a particle at Y as they are at different displacements and moving in different directions. There is half a cycle of difference and this can be given as 180° or π radians.

(d) (i) Periodic time is the time taken for one complete oscillation of the wave. ✓

(e) Periodic time is sometimes referred to as just period. When a definition is required, try to be as detailed as possible. It would not be acceptable to say that periodic time is for one oscillation.

(ii) $f = \dfrac{v}{\lambda}$ ✓

$f = \dfrac{12}{4} = 3\,\text{Hz}$ ✓

$T = \dfrac{1}{f} = \dfrac{1}{3} = 0.33\,\text{s}$ ✓

(e) The periodic time cannot be found directly from the displacement–distance graph. The speed of the wave has been given and the wavelength has been determined so frequency can be calculated and frequency is related to period. Make sure the equations you use are written in your answer. Check the following: equation, substitutions, answer, unit. Make sure that the units of speed and wavelength are consistent.

Question 2

(a) State Snell's law of refraction. (2 marks)

(b) Describe an experiment to determine the refractive index of glass in the form of a rectangular block. (8 marks)

(c) A ray of light is incident on a plane glass–air boundary, making an angle of 25°. The refractive index of glass is 1.50.

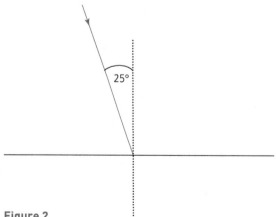

Figure 2

Complete Figure 2 to show the resultant paths of the rays of light. Make any necessary calculation to find the angles involved. (4 marks)

(d) Determine the angle of incidence at the glass–air boundary that would result in the angle of refraction becoming 90°. (2 marks)

(e) Describe the path taken by the ray of light when the angle of incidence is increased beyond the value calculated in (d). (2 marks)

Total: 18 marks

(a) Snell's law states that the ratio of the sine of the angle of incidence to the sine of the angle of refraction is the same ✓ for all rays travelling across a given boundary ✓.

e There are 2 marks for this question and you must earn both. Ideally, you will learn this definition thoroughly and state it exactly in words. The first mark is for stating that this specific ratio is a constant. This may be given as an equation but make sure all the terms are defined ($\sin i/\sin r$ = constant, where i is the angle of incidence and r is the angle of refraction). The second mark is for explaining that this ratio is for the boundary between two particular materials. For another pair of materials there is a different ratio. Note that n, the refractive index, is not mentioned in the law.

(b)

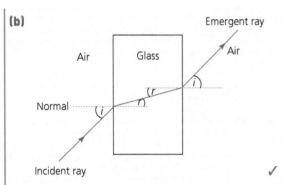

Trace around the block, remove the block ✓, draw a normal and measure and mark five angles of incidence. ✓

Replace the block and direct the ray of light along one of the incident paths and mark the emergent ray for each angle of incidence. ✓

Remove the block and join the incident and emergent rays. ✓ Measure the angle of refraction in the block with a protractor. ✓

Plot a graph of $\sin i$ against $\sin r$, draw a best-fit straight line ✓, the gradient of which is the refractive index ✓.

e In most examination papers you will be asked to describe an experiment that is included in the specification. You will normally be given a space in which you can draw a labelled diagram. Draw the diagram with a ruler. Describe the experiment in detail stating what measurements you will make, how you will make them and what you will do with the measurements to lead to your conclusion.

(c)

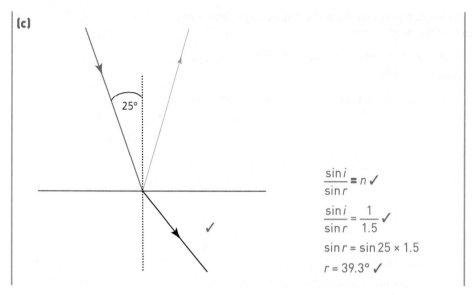

$$\frac{\sin i}{\sin r} = n \checkmark$$

$$\frac{\sin i}{\sin r} = \frac{1}{1.5} \checkmark$$

$$\sin r = \sin 25 \times 1.5$$

$$r = 39.3° \checkmark$$

ⓔ When you complete diagrams to show reflection or refraction, always draw a normal. This will help you to make the correct measurements and demonstrate clearly the correct path of the light. As the exact angles can be calculated in this question, it is important to show these angles accurately in the diagram. In questions that involve refraction, check carefully whether the light is travelling from a low-density material to a high-density material or from a high-density material to a low-density material. In this question it is travelling from glass to air so the light will bend away from the normal. The question tells us the refractive index for glass, $_an_g = 1.5$, but this is for light going from air to glass. We need the refractive index from glass to air, $_gn_a = 1/1.5$. This is a commonly made mistake. The first mark is a 'soft' mark for quoting a 'recall' equation.

(d) $\sin c = \dfrac{1}{n}$

$\sin c = \dfrac{1}{1.5} \checkmark \Rightarrow c = 41.8° \checkmark$

ⓔ The first mark is for recognising that the critical angle is required, or for the substitutions into the equation $\sin i/\sin r = n$ for an angle of refraction of 90°. The critical angle of glass and Perspex is approximately 42°. Try to remember this value as a check on your answer.

(e) Total internal reflection will occur ✓ and the light will be reflected back into the glass with the angle of reflection equal to the angle of incidence ✓.

ⓔ There are 2 marks for this answer, so two distinct aspects are required. The first mark is for simply stating that total internal reflection takes place. The second is awarded for more detail of this phenomenon. Always be aware of the mark allocation.

Question 3

(a) Complete Figure 3 to show where parallel rays of light from a distant object would be focused by an eye of a person suffering from myopia. Explain your completed diagram. (3 marks)

Figure 3

ⓔ This question examines your understanding of lenses and defective eyesight. First, identify which sight defect is being tested. In this question it is short sight — myopia.

(b) Complete the Figure 4 to show the type of lens used to correct myopia and show where the rays are now focused. Explain your completed diagram. (2 marks)

Figure 4

(c) What component of the eye provides the greatest refracting power in the eye? (1 mark)

(d) What is meant by the term accommodation in the context of eyesight? (1 mark)

(e) Describe the mechanism of this process and explain why a malfunction of this process may lead to myopia and hypermetropia. (3 marks)

Total: 10 marks

(a)

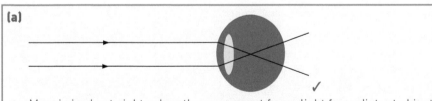

Myopia is short sight, when the eye cannot focus light from distant object on the retina. ✓ The eye lens is too powerful and rays meet before retina. ✓

ⓔ Refraction takes place as the light enters the eye at the air–cornea boundary and at the lens. If possible show this on the diagram but as long as you show the overall converging of the rays that will suffice. Clearly show the rays crossing before the retina.

(b)

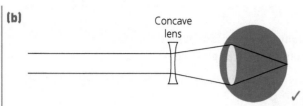

Concave lens

A concave (diverging) lens is used to spread the rays further apart, giving the eye lens more work to do and so bring the rays to a point on the retina. ✓

ⓔ Make sure you learn that concave lenses help myopia and convex lenses help hypermetropia.

(c) The boundary between air and the cornea has the greatest difference in optical density and therefore the greatest refractive index. Hence the greatest amount of refraction takes place here in the eye, but this is a fixed amount of refraction. ✓

ⓔ It is not the eye lens. Always give as full an answer as possible.

(d) The ability of the eye to produce clear images of objects over a wide range of distances from the eye. ✓

ⓔ It is not just the ability to produce images. You must refer to clear or sharp images.

(e) The ciliary muscles in the eye stretch or squash the shape of the lens to make it thicker, with a short focal length, to view near objects clearly and thinner, with a longer focal length, to view far objects clearly. ✓

If the eye cannot make the lens sufficiently thick then the focal length cannot be made short enough to view close objects as the lens is not as powerful as it needs to be to bring the rays to a focus on the back of the eyeball and instead they cross after the retina. This is hypermetropia or long sight. ✓

If the eye cannot make the lens sufficiently thin then the focal length cannot be made long enough to view distant objects and the lens is too powerful and brings the rays to a focus before the retina. This is myopia or short sight. ✓

ⓔ Don't confuse myopia and hypermetropia.

Question 4

Sea waves enter a harbour and are diffracted as they pass through the harbour gate (Figure 5).

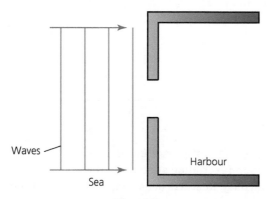

Figure 5

(a) Explain the meaning of diffraction. (2 marks)

(b) Describe the conditions necessary for observable diffraction. (1 mark)

(c) Add three more wavefronts to Figure 5 to illustrate the waves entering the harbour. (2 marks)

(d) At another part of the coast the sea waves change direction and their wavelength decreases. Why does this happen? (2 marks)

Total: 7 marks

(a) Diffraction is the spreading of waves ✓ when they pass through an opening or round an obstacle ✓.

e Learn this definition of diffraction.

(b) For observable diffraction the width of the gap through which the wave is passing must be of approximately the same size as the wavelength of the wave. ✓

(c)

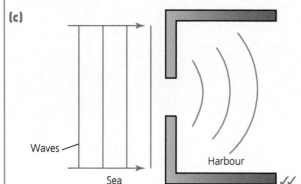

ℯ A mark will be awarded for the three wavefronts that have been requested if they are the correct shape. The second mark is for accuracy of your diagram, which must show that there is no change in wavelength, and you must indicate this by maintaining a constant distance between the wavefronts. It is likely that a mark would be deducted if only two and not three wavefronts were drawn.

(d) Change of direction and decreasing wavelength mean the wave has slowed down ✓ and has been refracted, so it must have entered shallower water at an angle ✓.

Question 5

(a) Describe an experiment to measure the wavelength of a laser using a diffraction grating of known line spacing. Your description should use the headings: Apparatus, Procedure, Safety warning, Result processing. (8 marks)

ℯ This experiment is listed in the specification and, as such, should have been performed or demonstrated in class and written up.

(b) (i) A certain diffraction grating has a series of opaque lines of width 5.5×10^{-4} mm separated by spaces of width 7.0×10^{-4} mm. Calculate the number of lines per millimetre on the grating. (2 marks)

(ii) A beam of monochromatic light incident normally on the grating produces a number of visible beams on the opposite side of the grating. One of the beams passes straight through the grating. Two of the other beams, symmetrically placed about the undeviated beam, are separated by an angle of 98.0°. Calculate the wavelength of the light used. (5 marks)

Total: 15 marks

(a) Apparatus: laser, grating, screen shown and labelled. ✓

Procedure:
- Assemble the apparatus as shown and turn off the room lights/darken the room. ✓
- Measure the distance from grating to screen, D. ✓
- Observe different orders of the diffracted beam and measure the distance, s, from the central beam ($n = 0$) to each of the diffracted beams. ✓

Safety warning: Do not look directly into the laser light source. ✓

Result processing:
- Calculate the angle, θ, for each order observed ($\tan \theta = s/D$). ✓
- Use the equation $n\lambda = d \sin \theta$ for each corresponding pair of n and θ results. ✓
- Calculate a series of λ values and take an average. ✓

ℯ You should have written up exactly how the experiment was performed in class. Recalling the procedure sequentially will prevent detail being overlooked and marks lost.

(b) (i) $d = (5.5 + 7.0) \times 10^{-4} = 12.5 \times 10^{-4}$ mm ✓

lines per mm $= \dfrac{1}{12.5 \times 10^{-4}} = 800$ lines per mm ✓

(ii) $d \sin \theta = n\lambda$

Therefore

$12.5 \times 10^{-7} \times \sin 49 = n\lambda$ ✓

9.43×10^{-7} m $= n\lambda$ ✓

visible wavelength range $= 400{-}700$ nm ✓

If $n = 1$, $\lambda = 943$ nm — too long for visible; $n = 3$, $\lambda = 314$ nm — too short ✓; $n = 2$, $\lambda = 472$ nm — within the visible range, therefore selected ✓.

ⓔ Watch out for the required conversion from mm to m, to avoid the classic $10n$ error.

Question 6

(a) (i) A lens produces an image of an object. Define the resulting **linear magnification**. (1 mark)

(ii) An object is placed on the principal axis of a thin lens of focal length f, at a distance u from the centre of the lens. An image of the object is formed at a distance v from the lens. Which of the following ratios is/are equal to the linear magnification produced by the lens?

$\dfrac{u}{v} \qquad \dfrac{u}{f} \qquad \dfrac{v}{u} \qquad \dfrac{v}{f}$ (1 mark)

(b) (i) A thin converging lens has a focal length of 200 mm. Find, by calculation, at what distance from the lens an object must be placed, on the principal axis of this lens, to obtain a real image with linear magnification equal to 3. (4 marks)

(ii) If the image remains real, in which direction, along the principal axis, must the object in (b)(i) be moved to reduce the linear magnification to a value less than 3? (1 mark)

(iii) How far must the object in (b)(i) be moved along the principal axis to obtain a real image equal in size to the object? (3 marks)

Total: 10 marks

ⓔ A good working knowledge of ray diagrams is invaluable when doing these calculations.

(a) (i) The ratio of the height of the image to the height of the object. ✓

(ii) $\dfrac{v}{u}$ ✓

(b) (i) $3 = \dfrac{v}{u}$

Therefore $v = 3u$ ✓

Using $\dfrac{1}{u} + \dfrac{1}{v} = \dfrac{1}{f}$:

$$\frac{1}{u} + \frac{1}{3u} = \frac{1}{200} \checkmark$$

Therefore $\frac{3}{3u} + \frac{1}{3u} = \frac{1}{200}$ and $\frac{4}{3u} = \frac{1}{200}$ $\checkmark$

$3u = 800$

Therefore $u = 267\,\text{mm}$ $\checkmark$

ⓔ Beware of wrong mathematics, for example '$1/u + 1/3u = 1/200$, so $4u = 200$'. To check your answer, remember that $u > f$ results in a real image, while $u < 2f$ gives a magnified image.

(ii) Away from the lens. $\checkmark$

(iii) The object and image are the same size when $u = v = 2f$, in this case 400 mm. $\checkmark$

Therefore it should be moved from 267 mm to 400 mm. $\checkmark$

$\Delta u = 133\,\text{mm}$ $\checkmark$

Question 7

You are to determine the speed of sound in air by the resonance tube method.

(a) Draw a labelled sketch of the experimental arrangement you would use. (2 marks)

(b) What procedure would you follow to ensure that the *first* position of resonance is used on each occasion? (2 marks)

(c) What measurements would you take, and how would you use then to obtain a reliable value for the speed of sound in air? (3 marks)

(d) A standing wave has been set up in the resonance tube. State which physics principle is involved and describe how the standing wave is formed in the tube. (3 marks)

Total: 10 marks

(a) (Clear) resonance tube (open at both ends) in a water container, with metre rule alongside. $\checkmark$

Loudspeaker with signal generator or tuning fork , just above the open end of the pipe. $\checkmark$

(b) For a frequency, begin with zero length of air column and increase *or* for a fixed length, begin at the lowest frequency and increase $\checkmark$; until a distinct increase in loudness is detected $\checkmark$.

(c) Record frequency and length of air column at resonance. $\checkmark$

Repeat for various frequency or length values. $\checkmark$

Calculate $v = 4fl$ and average *or* plot frequency against 1/length and calculate $v = 4 \times$ gradient. $\checkmark$

(d) Superposition ✓

> Two waves travelling in opposite directions — the incident wave and the reflected wave at the water surface. ✓
>
> Resonance occurs when driving frequency = natural frequency. ✓

e The word 'reliable' in the text triggers the need to have more than one set of results to determine a quantity. By having several readings, anomalies can be recognised and disregarded.

Question 8

Some of the energy levels for the mercury atom are –3.70 eV, –5.51 eV and –10.40 eV (ground state).

(a) Calculate the energy required to excite an electron in the ground state to each of the other two levels. (2 marks)

(b) Describe what would happen if cold mercury is bombarded with:
   (i)   electrons of kinetic energy 2.5 eV (2 marks)
   (ii) electrons of kinetic energy 6 eV (2 marks)
   (iii) light of wavelength 254 nm (3 marks)
   (iv) light of wavelength 200 nm (3 marks)

Total: 12 marks

(a) It would take (–5.51 eV) – (–10.40 eV) = 4.89 eV to raise an electron from the ground state to the –5.51 eV energy level. ✓

It would take (–3.70 eV) – (–10.40 eV) = 6.70 eV to raise an electron from the ground state to the –3.70 eV energy level. ✓

(b) (i) The atom needs at least 4.89 eV to raise it from the ground state to a higher level. 2.5 eV is insufficient. ✓
The bombarding electrons collide with the electrons in the atoms, not giving up any of their energy, and move on. ✓

(ii) The atom needs at least 4.89 eV to raise it from the ground state to a higher level. 6 eV is more than sufficient. ✓
So the electron gives up 4.89 eV of energy and comes away with 1.11 eV. The excited atom will return to the ground state with the emission of a photon of energy 4.89 eV. ✓

(iii) Photons can only give up all or nothing of their energy.
A wavelength of 253 nm can be converted to energy using:

$$E = \frac{hc}{\lambda} = \frac{6.63 \times 10^{-34} \times 3 \times 10^{8}}{254 \times 10^{-9}}$$

$$= 7.83 \times 10^{-19}\,J = \frac{7.83 \times 10^{-19}}{1.6 \times 10^{-19}}\,eV = 4.89\,eV \checkmark$$

This is exactly the correct amount of energy so it is the exact wavelength needed and the photon is absorbed. ✓ The atom returns to the ground state with the release of a photon of 4.89 eV ✓.

(iv) $E = \dfrac{hc}{\lambda} = \dfrac{6.63 \times 10^{-34} \times 3 \times 10^8}{200 \times 10^{-9}}$

$= 9.945 \times 10^{-19}\,J = \dfrac{9.945 \times 10^{-19}}{1.6 \times 10^{-19}} = 6.22\,eV$ ✓

This is too much energy to promote an electron from the ground state to the first energy level and too little to promote an electron form the ground state to the second level. ✓

A photon must transfer all or none of its energy so the photons are not absorbed. ✓

ⓔ It is important to understand the difference between a photon providing the energy to electrons and the electrons receiving it through collisions with other electrons (or by heat). Photons are packets of electromagnetic energy and only exist as whole entities — they cannot give up part of their energy, just all or nothing.

Question 9

(a) With the aid of a labelled sketch, describe briefly how the wave properties of electrons can be demonstrated. (4 marks)

(b) Calculate the de Broglie wavelength of:
(i) a ball of mass 55 g travelling at 55 m s^{-1} (2 marks)
(ii) an electron with kinetic energy 105 eV. (4 marks)

Total for this self-assessment test: 100 marks   Total: 10 marks

(a) Diagram showing:
- metal foil or carbon film ✓
- fluorescent screen ✓
- diffraction pattern, circles ✓

Wording to describe fast moving electrons incident onto the target ✓

ⓔ The questions states 'with the aid of', so there should be a labelled sketch *and* wording. It is possible that a well-labelled sketch could gain full marks. The sketch should fill any space provided on the paper, and should be clear and well labelled. Don't spend too much time on it — this is not an art exam.

(b) (i) $\lambda = \dfrac{h}{p} = \dfrac{6.63 \times 10^{-34}}{0.055 \times 55}$ ✓

$\lambda = 2.2 \times 10^{-34}\,m$ ✓

(ii) K.E. $= 105 \times (1.6 \times 10^{-19}) = 1.68 \times 10^{-17}\,J$ ✓

K.E. $= \frac{1}{2}mv^2 = \frac{1}{2}\dfrac{p}{m}$

Therefore $p = \sqrt{(2m \times \text{K.E.})}$ ✓

$p = \sqrt{(2 \times [9.1 \times 10^{-31}] \times [1.68 \times 10^{-17}])} = 5.5 \times 10^{-24}\,\text{kg m s}^{-1}$ ✓

$\lambda = \dfrac{h}{p} = \dfrac{6.63 \times 10^{-34}}{5.5 \times 10^{-24}} = 1.2 \times 10^{-10}\,\text{m}$ ✓

■ Self-assessment test 2

Question 1

(a) List three properties common to all electromagnetic waves. (3 marks)

(b) Explain why the electromagnetic spectrum is described as continuous. (1 mark)

(c) State four electromagnetic waves that have shorter wavelengths than indigo light. (2 marks)

(d) Calculate the wavelength of radio waves of frequency 900 kHz. (3 marks)

(e) Describe and explain how light can be plane-polarised and suggest a method to check polarisation has taken place. (4 marks)

Total: 13 marks

(a) Electromagnetic waves:
- travel at the speed of light ✓
- are transverse waves ✓
- can travel through a vacuum ✓
- (or consist of oscillating electric and magnetic fields ✓)

ⓔ A frequent mistake is suggesting that electromagnetic waves are longitudinal waves. This is totally incorrect.

(b) There is a wave at every wavelength — there are no gaps. The waves overlap as there are no clear boundaries between the waves. ✓

(c) Violet light, ultraviolet radiation, X-rays, gamma rays ✓✓ (½ each)

ⓔ Know the electromagnetic spectrum in order of increasing wavelength and be able to suggest typical wavelengths for each wave in the spectrum.

(d) $f = 900\,\text{kHz} = 9 \times 10^5\,\text{Hz}$ ✓

$c = 3.0 \times 10^8\,\text{m s}^{-1}$

$v = f\lambda$ ✓

For electromagnetic waves $c = f\lambda$

$\lambda = \dfrac{c}{f} = \dfrac{3 \times 10^8}{9 \times 10^5} = 333\,\text{m}$ ✓

ⓔ Check the units to convert from multiples such as kilohertz. Write the formula in the correct form (changing the subject if necessary) before making substitutions.

(e) In unpolarised light the oscillations occur in all planes. ✓

When unpolarised light is passed through a polarising filter such as Polaroid the oscillations in all planes but one will be absorbed and the oscillations will be in one plane only. ✓

A second polariser is used as an analyser to confirm that the light is plane polarised. ✓

The analyser is rotated through 360° and the light intensity will be extinguished twice in the rotation. ✓

ⓔ Describe experiments or practical procedures carefully, with all the details. For example, don't just say an analyser is used to check polarisation. Describe in detail *how* it is used.

Question 2

Figure 1 shows a ray of light incident at an angle of 75° on a glass block. The ray enters the glass and meets the side BC. The critical angle for glass is 41°.

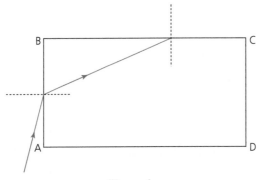

Figure 1

(a) (i) Explain what is meant by the term refraction of light. (1 mark)

    (ii) Explain what is meant by the term refractive index. (2 marks)

(b) Calculate the refractive index of the glass. (2 marks)

(c) Show that the ray will not emerge from the block at side BC. (3 marks)

Total: 8 marks

(a) (i) Refraction is the bending of light as it travels between materials of different densities and changes speed. ✓

(ii) Refractive index is a measure of the change of direction of light as it travels from air (or a vacuum) to another transparent material. ✓

It is the ratio of the speed of the light in air (or a vacuum) to the speed of the light in the second medium. ✓

(Or it is the ratio of the sine of the angle of incidence to the sine of the angle of refraction across the boundary between air (or a vacuum) and the second medium. ✓)

ⓔ To be awarded 2 marks a full description is required.

(b) $n = \dfrac{1}{\sin C}$

$n = \dfrac{1}{\sin 41}$ ✓

$n = 1.52$ ✓

(c) Find the angle of refraction:

$n = \dfrac{\sin i}{\sin r}$

$\sin r = \dfrac{\sin 75}{1.52}$

$r = 39.5$ ✓

Use this to find the angle of incidence at BC by considering the geometry of the block.

angle of incidence at BC = 50.5° ✓

This is greater than the critical angle so total internal reflection will take place and the light will not emerge from the block. ✓

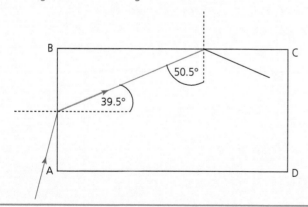

ⓔ It is not necessary to measure angles accurately with a protractor in this question. However, as you calculate the angles using Snell's relationship and the geometry of a triangle, mark the angles on the diagram and make sure they are a good approximation to the actual value. Explain your reasoning as you work through the question and show all working out.

Question 3

(a) Optical fibres are used in medicine in an instrument called an endoscope.

   (i) What is an endoscope used for?   (1 mark)

   The instrument contains two separate bundles of optical fibres.

   (ii) What are their functions?   (2 marks)

   (iii) How are the fibres arranged in each bundle?   (1 mark)

(b) The maximum time taken for a pulse of light to travel along an optical fibre is when it zig-zags along the fibre making an angle of incidence equal to the critical angle whenever it is reflected.

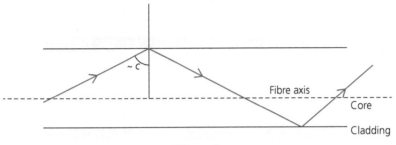

Figure 2

   The minimum time is when it travels along the axis of the fibre. If the critical angle for one such fibre is 88.0° and the refractive index of the core is 1.520, find how long the fibre can be if the difference in the time taken by the two routes is not to be greater than 2 ns.   (6 marks)

Total: 10 marks

ℯ The time difference relates to the signal dispersion of a light pulse on transmission. In effect, the duration of the pulse is increased. As a consequence, in a long communication fibre separate pulses can overlap and errors and loss of information will occur at the receiving end.

(a) (i) To view an internal part of the body. ✓

   (ii) One bundle is used to carry light into the body ✓, the other carries the image out of the body ✓.

   (iii) Light in — non-coherent; image out — coherent ✓

(b) $n = \dfrac{c_{air}}{c_{core}}$ ✓

   $c_{core} = \dfrac{3.00 \times 10^8}{1.520} = 1.97 \times 10^8 \, \text{m s}^{-1}$ ✓

   Let the length of the fibre = d.

   $t_{min} = \dfrac{d}{1.97 \times 10^8} \, \text{s}$ ✓

   The light following the zig-zag route will travel along the 'hypotenuse'.

$$t_{max} = \frac{d}{\sin 88.0°} \times \frac{1}{1.97 \times 10^8} \checkmark$$

$$\Delta t = t_{max} - t_{min}$$

$$\Delta t = 2 \times 10^{-9} = \frac{d}{1.97 \times 10^8} \times \frac{1}{\sin 88.0° - 1} \checkmark$$

$$d = 646\,m \checkmark$$

ⓔ The critical angle value is used even though, being precise, the reflected ray would travel along the core/cladding interface. 'Just below' the critical angle is what is required.

Question 4

(a) (i)   Using a graph page, draw a ray diagram for a converging lens. Label the principal axis, refracting plane of the lens and the principal foci. Draw an upright arrow to represent the object at a suitable position relative to your principal focus so that a real, magnified image of the object is formed. (3 marks)

   (ii) Make accurate measurements from your ray diagram to confirm that magnification $m = v/u$.

   Indicate clearly on your diagram the measurements you have taken. Write down their values and show how they are used in your confirmation of the formula. (3 marks)

(b) State the meaning of hypermetropia. (1 mark)

(c) State the type of lens used to correct hypermetropia. (1 mark)

(d) (i)   A person with hypermetropia has a near point of 600 mm. Complete Figure 3 to show how light from the person's near point would be focused by their unaided eye. (2 marks)

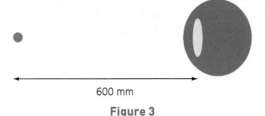

600 mm

Figure 3

   (ii) Calculate the power of the lens required to correct the person's near point to a normal distance. (3 marks)

Total: 13 marks

(a) (i)   Axis, foci, object placed between F and 2F ✓
        Two correct rays ✓✓

   (ii) Measured values of h_i and h_o ✓
        Measured values of v and u, as image and object distances ✓
        Evaluate h_i/h_o and equate to the value of v/u. ✓

(b) Distant objects can be focused but close objects cannot. ✓

(c) Convex — converging lens ✓

(d) (i)

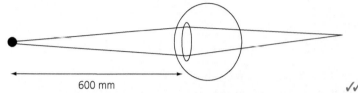

600 mm

✓✓

(ii) A normal near point is 25 cm. The lens will provide a virtual image at the actual near point of 600 mm when an object is placed at the corrected near point of 25 cm.

$$\frac{1}{f} = \frac{1}{u} + \frac{1}{v}$$

$$\frac{1}{f} = \frac{1}{250} + \frac{1}{-600} \checkmark$$

$$f = 429 \text{ mm} = 0.43 \text{ m} \checkmark$$

$$P = \frac{1}{f} = \frac{1}{0.43} = 2.3 \text{ D} \checkmark$$

ⓔ Remember to convert all quantities to the same units. Note that the question asks for the power and if dioptres is the unit then metres must be used for the focal length.

Question 5

Figure 4 represents two identical, air-filled pipes, each closed at one end. A vibrating tuning fork of frequency f_1, placed near the open end of one of the pipes, excites the first mode of vibration (the fundamental mode) of the air in this pipe. A second fork of frequency f_2 excites the next mode of vibration (the first overtone) of the air in the other pipe. The points P, Q, R and S in the diagram are equally spaced along the lengths of the pipes.

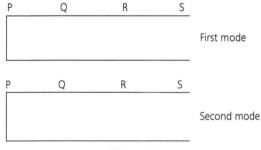

Figure 4

(a) Complete Figure 4 to illustrate the wave pattern representing the mode of vibration in each pipe. On your diagram label clearly the positions of all nodes with the letter N, and the positions of all antinodes with the letter A. (5 marks)

(b) State the value of the ratio of frequencies of the forks, f_1/f_2. (1 mark)

(c) Given that the speed of sound in the air in the pipes is 330 m s⁻¹, and the difference in frequency between the two forks is 330 Hz, calculate the length of the pipes. (4 marks)

(d) Consider the *first* stationary mode of vibration of the air in the pipe. Sketch graphs to show how the relative amplitudes of vibration of the air disturbance in the pipe, at positions P, Q, R and S along the pipe, vary with time. (4 marks)

Total: 14 marks

(a)

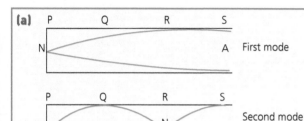

Correct first mode ✓
Correct second mode ✓
Nodes correctly marked ✓
Antinodes correctly marked ✓
Correct number of nodes and antinodes ✓

(b) $\dfrac{f_1}{f_2} = \dfrac{1}{3}$ ✓

(c) $f_2 - f_1 = 330$ and $f_2 = 3f_1$ ✓

Therefore:

$3f_1 - f_1 = 330$ and $f_1 = 165$ Hz ✓ (Consequently $f_2 = 495$ Hz)

$\lambda = \dfrac{v}{f} = \dfrac{330}{165} = 2.0$ ✓

$\lambda_1 = 4 \times$ length of pipe length $\Rightarrow$ of pipe = 0.50 m ✓

(d)

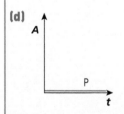 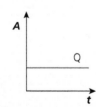 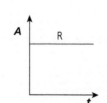 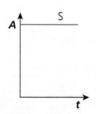

Correctly labelled axes ✓
Use of amplitude rather than displacement ✓
Horizontal lines ✓
Correct relative magnitudes ✓

ⓔ This is tricky, but can be accessed by considering the definitions of the node and antinode — always zero displacement and periodic maximum displacement, respectively. Then the intermediary positions fall in between.

Question 6

(a) What are the basic principles of the production of X-rays?   (6 marks)

(b) In what way does a CT image differ from a conventional X-ray image?   (4 marks)

(c) State why all X-ray imaging is viewed as harmful.   (1 mark)

Total: 11 marks

(a) Electrons from a hot cathode ✓ are accelerated by a high voltage ✓ in an evacuated chamber ✓ to a target metal anode ✓.

Higher-level electrons drop energy levels ✓ to fill vacancies left by electrons in inner shells being knocked out by the incident electrons ✓.

ⓔ This is a complicated concept. It is helpful when you are revising to learn to draw and label the X-ray tube and practise explaining the function of each part.

(b) The CT image is produced by a rotating X-ray tube and range of detectors that take multiple images of slices of part of the body ✓, which are combined by a computer to give a three-dimensional image ✓.

The conventional X-ray is produced by a stationary X-ray tube and a detector ✓, which produce a two-dimensional image of part of the body. ✓

(c) X-rays are ionising radiation and can damage living cells. ✓

ⓔ Comparison of conventional two-dimensional and three-dimensional X-rays is a common question. A comparison of the safety of the two types should prompt a discussion of the increased amount of dangerous ionising X-rays received in CT scans.

Question 7

Figure 4 shows the ground state and some of the excited energy levels in the hydrogen atom.

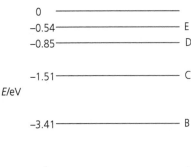

Figure 4

(a) To excite an electron from the ground state to a higher level requires energy. Where might this energy come from? (3 marks)

(b) How much energy is required to move an electron from the ground state out of the atom? (1 mark)

(c) Why do the energy levels have negative energy levels? (2 marks)

(d) What type of spectrum results from transitions of the electron between energy levels in the atom? (1 marks)

(e) Calculate the wavelength of radiation emitted by a transition between level E and level B and state the region of the electromagnetic spectrum corresponding to this wavelength. (3 marks)

(f) Laser action depends on the stimulated emission of radiation. Explain in terms of energy levels, what is meant by the term 'stimulated emission'. (3 marks)

Total: 13 marks

(a) The electrons can gain energy by heating, by collision or by absorption of photon energy. ✓✓✓

(b) 13.6 eV ✓

(c) The energy of an electron just outside the atom is taken as zero ✓ and if energy is added to get the electron out of the atom then the energy levels must be negative. ✓

(d) Line spectrum ✓

(e) change in energy $= (-0.54\,\text{eV}) - (-3.41\,\text{eV}) = 2.87\,\text{eV} = 2.87 \times 1.6 \times 10^{-19}\,\text{J}$ ✓ $= \dfrac{hc}{\lambda}$
$\lambda = \dfrac{6.63 \times 10^{-34} \times 3 \times 10^8}{2.87 \times 1.6 \times 10^{-19}} = 4.33 \times 10^{-7}\,\text{m} = 433\,\text{nm}$ ✓ = visible light ✓

ℯ Remember it is E_2 (the higher energy level) – E_1 (the lower energy level) = the energy of the emitted photon. Keep the minus signs in the correct positions to obtain the correct result. Always convert from eV to joules to use formulae.

(f) The atom is initially in an excited state, with electrons in higher energy levels. ✓

An incident photon of exactly the right energy causes an electron to fall to a lower level ✓ with the emission of a photon of the same energy and in phase with the incident one ✓.

ℯ This part of the question is only awarded 3 marks, so only a brief description is required, focused on the energy levels. Another question with a higher mark allocation might require a more detailed description.

Question 8

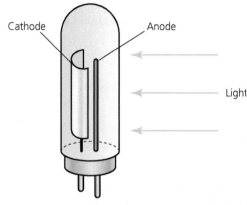

Figure 5

A photocell (Figure 5) is a device that can be used to measure the intensity of light incident upon it. When photons of light hit the cathode, some electrons may be released. This emission depends on the energy of the photon and the work function of the cathode.

(a) Explain the meaning of the term work function and state the conditions under which photoelectric emission will take place. (2 marks)

(b) Caesium has a work function of 1.9 eV. Calculate the lowest frequency of radiation that will produce photoelectric emission from a cathode containing caesium. (3 marks)

(c) If red light shines on the cathode, will this cause photoelectric emission? (3 marks)

Total: 8 marks

(a) The work function of a material is the minimum energy required to remove an electron from the surface of that material. ✓

Photoelectric emission will take place if electromagnetic radiation incident on the surface of the material provides photons of energy greater than or equal to the work function. ✓

ⓔ This question is about the work function of a metal. Do not answer in terms of threshold frequency.

(b) Convert eV to joules.

$1.9\,\text{eV} = 1.9 \times 1.6 \times 10^{-19} = 3.04 \times 10^{-19}\,\text{J}$ ✓

The work function is given by Planck's constant times the threshold frequency

$\varphi = hf_0$

So:

$$f_0 = \frac{\varphi}{h} = \frac{3.04 \times 10^{-19}}{6.63 \times 10^{-34}} \checkmark$$

$$f_o = 4.6 \times 10^{14}\,\text{Hz} \checkmark$$

(c) Red light has a wavelength of about 700 nm. ✓

So the frequency of red light is:

$$f = \frac{c}{\lambda} = \frac{3.0 \times 10^8}{700 \times 10^{-9}} = 4.3 \times 10^{14}\,\text{Hz} \checkmark$$

This is not a high enough frequency to cause photoemission. ✓

ⓔ Be able to suggest typical wavelengths for all electromagnetic waves including all the visible light waves.

Question 9

(a) What is meant by 'Doppler red shift' and how does it support the theory of an expanding universe? (3 marks)

(b) The quasar 3C273 has a prominent emission line at a wavelength of 502 nm. The wavelength of the same line measured from the hydrogen spectrum of a stationary source on Earth is 434 nm.

  (i) Calculate the recessional speed of 3C273. (2 marks)

  (ii) Taking the value of the Hubble constant (H_0) as $2.4 \times 10^{-18}\,\text{s}^{-1}$, estimate the distance of 3C273 from Earth (give your answer in light years). (2 marks)

(c) By consideration of the conservation of energy, would it be expected that the universe will continue to expand or eventually slow down? (3 marks)

Total for this self-assessment test: 100 marks   Total: 10 marks

(a) Doppler red shift is the change in wavelength of a spectral line towards the red end of the visible spectrum ✓ as the source moves away from the observer ✓.

The spectra from distant galaxies show red shift, so they must be moving away and the universe is expanding. ✓

(b) (i) $z = \dfrac{\Delta\lambda}{\lambda} = \dfrac{v}{c}$

$$z = \frac{\Delta\lambda}{\lambda} = \frac{502 - 434}{434} = 0.157 \checkmark$$

$$v = z \times c = 0.157 \times (3.00 \times 10^8) = 4.70 \times 10^7\,\text{m s}^{-1} \checkmark$$

ⓔ As z is a ratio, both $\Delta\lambda$ and λ can remain in nm and do not have to be converted to m.

(ii) $v = H_0 d \Rightarrow d = \dfrac{v}{H_0}$

$d = \dfrac{4.7 \times 10^7}{2.4 \times 10^{-18}} = 2.0 \times 10^{25}\,\text{m}$ ✓

$1\,\text{ly} = 9.46 \times 10^{15}\,\text{m} \Rightarrow d = 2 \times 10^9\,\text{ly}$ ✓

ℯ You are asked for an estimate of the distance, as the Hubble constant value is not known exactly, so the answer has been quoted to 1 s.f.

(c) The universe is gaining gravitational potential energy as it expands ✓, and so will lose kinetic energy ✓, causing the expansion to slow down ✓.

Knowledge check answers

1 In transverse waves the vibrations are at right angles to the direction of travel of the wave. Examples include any electromagnetic wave and water waves. In longitudinal waves the vibrations are parallel to the direction of travel of the wave. Examples include sound, ultrasound and seismic p-waves.

2 a Frequency can be determined from a displacement–time graph. The time between successive crests will give the periodic time. This can be used to find the frequency, as periodic time is $\dfrac{1}{\text{frequency}}$ (or $f = \dfrac{1}{T}$).

b Frequency cannot be determined from a displacement–distance graph. Wavelength is taken as the distance between successive crests and if the speed of the wave is known only then can the frequency be determined by using the wave equation.

3 $T = 1\,\text{ms} = 0.001\,\text{s}$ (change milliseconds to seconds)
$$\text{frequency} = \frac{1}{\text{period}} = \frac{1}{0.001} = 1000\,\text{Hz}$$

4 The wave equation is $v = f\lambda$.
$v = 68\,000 \times 0.005 = 340\,\text{m\,s}^{-1}$

5 phase difference $= \dfrac{3.5}{4} \times 360° = 90°$ or $\dfrac{\pi}{2}$ radians

6 The light would vary gradually from a maximum to zero intensity twice in the rotation.

7 Radio waves travel at the speed of light, can travel through a vacuum and are transverse waves. Radio waves are not considered as dangerous as they have a long wavelength, low frequency or low energy, and are not ionising radiation.

8 The speed will increase, the wavelength will increase and the frequency will stay the same.

9 Proportionality — the graph is a straight line through the origin and as you double the dependent variable, the independent variable also doubles.

10 $n = 1.52 = \dfrac{\text{speed of light}}{\text{speed of light in glass}} = \dfrac{3.00 \times 10^8}{c_g}$
$c_g = 1.97 \times 10^8\,\text{m\,s}^{-1}$

11 Light is faster than electricity; there is less interference of signal; the raw materials are cheaper; more information can be carried.

12 The image must be transmitted with a coherent bundle because the fibres will maintain a constant position and so the image will be an accurate copy of the object. This is not required for illumination.

13 A point on the principal axis through which rays of light travelling parallel to the principal axis converge after refraction by the lens.

14 Erect (upright), enlarged (magnified), virtual (on the same side of the lens as the object).

15 $\dfrac{1}{f} = \dfrac{1}{4} + \dfrac{1}{-8} = \dfrac{1}{8}$ $f = 8\,\text{cm}$

16 The diverging lens is thin in the middle, the convex lenses are thick in the middle and the fatter convex lens has the shortest focal length.

17 $\dfrac{1}{v} = \dfrac{-1}{u} + \dfrac{1}{f}$

if $\dfrac{1}{v} = 0$ then $\dfrac{1}{u} = \dfrac{1}{f}$

if $\dfrac{1}{u} = 0$ then $\dfrac{1}{v} = \dfrac{1}{f}$

18 A powerful lens has a short focal length and is a thick lens.

19 The greatest amount of refraction in the eye takes place at the air–cornea boundary because there is a large difference in refractive index between the air and the cornea. Under water there is very little difference between the water and the cornea and less refraction takes place when light enters the eye. The eye lens is not sufficiently powerful to focus the light onto the retina itself.

20 Normal near point to normal far point: 25 cm to infinity.

21 Waves that meet must be coherent, have the same amplitude and be out of phase by 180°.

22 Standing — the wave does not transfer energy from one place to another.
Mechanical — the wave is generated by disturbances in a material medium.
Longitudinal — there are areas of compression and rarefaction.

23 The fringe separation increases.

24 Quantised refers to discrete units or whole packets of energy.

25 The tungsten target absorbs the electrons and releases some of the energy in the form of X-rays by the rapid deceleration of electrons after passing the nucleus or by the tightly bound inner electrons being knocked out of the atom by the incident electrons and higher-state electrons dropping down to fill the vacancy.

26 distance = velocity × time = $(3 \times 10^8) \times (365 \times 24 \times 60 \times 60)$
$= 9.46 \times 10^{15}\,\text{m}$

27 $1\,\text{Mpc} = 3.08 \times 10^{22}\,\text{m}$
$74\,\text{km\,s}^{-1}\,\text{Mpc}^{-1} = \dfrac{7.4 \times 10^4}{3.08 \times 10^{22}}\,\text{s}^{-1} = 2.4 \times 10^{-18}\,\text{s}^{-1}$

Index